Astrobiology: A Very Short Introduction

VERY SHORT INTRODUCTIONS are for anyone wanting a stimulating and accessible way in to a new subject. They are written by experts and have been translated into more than 40 different languages. The series began in 1995 and now covers a wide variety of topics in every discipline. The VSI library contains nearly 400 volumes—a Very Short Introduction to everything from Indian philosophy to psychology and American history—and continues to grow in every subject area.

Very Short Introductions available now:

David C. Catling

ASTROBIOLOGY

A Very Short Introduction

OXFORD
UNIVERSITY PRESS

OXFORD
UNIVERSITY PRESS

Great Clarendon Street, Oxford, OX2 6DP,
United Kingdom

Oxford University Press is a department of the University of Oxford.
It furthers the University's objective of excellence in research, scholarship,
and education by publishing worldwide. Oxford is a registered trade mark of
Oxford University Press in the UK and in certain other countries

First Edition published in 2013

Impression: 3

Published in the United States of America by Oxford University Press
198 Madison Avenue, New York, NY 10016, United States of America

British Library Cataloguing in Publication Data
Data available

Library of Congress Control Number: 2013940856

ISBN 978-0-19-958645-5

Printed in Great Britain by
Ashford Colour Press Ltd, Gosport, Hampshire

Contents

Acknowledgements

I thank Professor John Armstrong, Dr Rory Barnes, Professor
John Baross, Dr Billy Brazelton, Professor Roger Buick, Dr Rachel
Horak, Imelda Kirby, and Professor Woody Sullivan for reading
parts of the manuscript and offering various corrections and
suggestions. Thanks also to Latha Menon and Mimi Southwood,
who read the entire manuscript, the latter offering the perspective
of a non-scientist. Numerous students who attended my
astrobiology classes at the University of Washington over the years
also helped me mentally prepare for writing this book. At Oxford
University Press, I thank Emma Ma and Latha Menon for their
encouragement and assistance.

List of illustrations

Chapter 1
What is astrobiology?

Behind the name

'What the hell is *astrobiology*?' an American Secret Service agent cried into his walkie-talkie. He had just been checking the identity of an academic visitor to NASA's Ames Research Center, near San Francisco. The visitor had said that he was attending NASA's first astrobiology science conference. Ames has an airstrip that provides a secure landing site for Air Force One, and, in April 2000, President Bill Clinton had just flown in to visit the San Francisco Bay area, bringing along his Secret Service entourage.

The agent's question was a fair one. It was only in the late 1990s that a scientific consensus emerged about the meaning of *astrobiology*. Few laymen or Secret Service agents would have heard of the term. Back then, NASA began to promote a research programme in astrobiology led by Ames, where I was working as a space scientist. At first, some of my colleagues disliked the literal Greek meaning of the 'biology of stars'. One noted with a scoff how life couldn't exist inside the infernos of stars. A less curmudgeonly interpretation is that the 'astro' in astrobiology concerns life *around* stars, including the Sun, or simply life in space. In fact, many astrobiologists are as much concerned with the history of life on Earth as with life elsewhere. Astrobiologists agree that we should have a firm understanding of how life evolved on Earth in order to ponder the existence of life in outer space. Yet one of the astonishing aspects of modern science is that it has so far failed to answer questions about biology that even a child might ask. How did life on Earth get started? We have some ideas but the details are unknown. Which special properties of the Earth and the Solar System make our planet habitable? Again, some thoughts but there is still much to learn. And what caused life to evolve into complex organisms instead of remaining simple? Again, we're uncertain.

To fill these holes in human knowledge, astrobiology has emerged as *a branch of science concerned with the study of the origin and evolution of life on Earth and the possible variety of life elsewhere*. This is my own preferred definition. NASA has defined astrobiology as *the study of the origins, evolution, distribution, and future of life in the universe*. Other common definitions are *the study of life in the universe* or *the study of life in a cosmic context*. Within this purview, astrobiologists pursue the question 'What's the history and future of terrestrial life?' as well as 'Is there life elsewhere?'

Four developments coincided with the emergence of astrobiology as a discipline in the late 1990s. In 1996, controversial signs of ancient life were described within a Martian meteorite—a 1.9 kilogram piece of rock that had been blasted off the surface of Mars by an asteroid impact and had eventually landed in Antarctica. Whether the interpretation of fossilized microscopic life was correct or not (see Chapter 6), it set people thinking. Furthermore, over the preceding two decades, biologists had established that some microbes not only tolerated a much larger range of environments than had previously been thought but actually *thrived* in extremes of temperature, acid, pressure, or salinity. So it became plausible to contemplate extraterrestrial microbes existing in seemingly hostile places. A third finding came in 1996 from pictures taken by NASA's *Galileo* spacecraft of the ice-covered surface of Jupiter's moon, Europa, which revealed pieces of ice that had drifted apart in the past, suggesting an ocean below an icy crust. Then, from the mid 1990s onwards, astronomers found increasing numbers of extrasolar planets or *exoplanets*, which are planets orbiting not our Sun but other stars. The possibility that life might reside on exoplanets or in the cosmic backyard of our own Solar System provided an impetus for asking whether life might be common in the universe.

Astrobiology in the history of ideas

Although astrobiology came to the fore in the 1990s, the question of whether we're alone in the universe goes back millennia. Thales (*c*. 600 BC), often regarded as the father of Western philosophy, espoused the idea of a *plurality of worlds* with life. Subsequently, the Greek atomist school from Leucippus to Democritus and Epicurus, which believed that matter was made of indivisible atoms, favoured such 'pluralism'. Metrodorus (*c*. 400 BC), a follower of Democritus, wrote, 'It is unnatural in a large field to have only one stalk of wheat, and in the infinite universe only one living world'. But it would be incorrect to equate the ancient philosophers' pluralism with our modern conception of life on Mars or exoplanets. Metrodorus had no clue that stars were Sun-like objects at enormous distances and believed that they formed daily from moisture in the Earth's atmosphere. The populated worlds of the atomists' imagination were bodies in an intangible space, similar to modern ideas of parallel universes. In any case, the opposing view of Plato (427–347 BC) and Aristotle (384–322 BC) ultimately dominated. Their belief that the Earth was uniquely inhabited and at the centre of the universe prevailed for over a thousand years.

Eventually, Renaissance astronomers showed that the Earth orbited the Sun. With the realization that the Earth was merely another planet, speculations soon arose about extraterrestrial life on other planets in the Solar System. Johannes Kepler (1571–1630), the German astronomer responsible for astronomy's three laws of planetary motion, happily entertained the idea of inhabited planets. Then, by the end of the 17th century, the Dutch astronomer Christiaan Huygens (1629–95) was imagining life *beyond* the Solar System in his book *Cosmotheoros* (1698): 'all those Planets that surround that prodigious number of Suns. They must have their plants and animals, nay and their rational creatures too.' Extraterrestrial life was so in vogue that in 1755 the philosopher Immanuel Kant (1724–1804) wrote of intellectuals on Jupiter and amorous Venusians.

In contrast, some scholars with religious views continued to cling to Earth's uniqueness. An example was Cambridge University's William Whewell (1794–1866), whose *Of the Plurality of Worlds* (1853) argued against other inhabited planets in a sort of forerunner of a contemporary debate called the *Rare Earth Hypothesis* that I discuss in Chapter 8.

By the late 19th century, the issue of life elsewhere was seen as a purely scientific matter, though science itself soon developed some blind alleys. Telescopic observations by Giovanni Schiaparelli (1835–1910) and Percival Lowell (1855–1916) created a surge of interest in the possibility of intelligent life on Mars. Unfortunately, Lowell's belief that he saw canals on Mars was an optical illusion created when the mind connects dots in blurry images, and his ideas of Martian civilizations were fantasy. Increasingly, as astronomers employed painstaking techniques such as examining spectra of light from planets, it became apparent that the physical conditions on various Solar System planets might not be so favourable for life after all. The pendulum swung so firmly in the other direction that by the mid 20th century few astronomers were interested in planets. It took the Space Age to rekindle old curiosities.

While the basic questions of astrobiology are ancient, the term 'astrobiology' only surfaced from time to time before becoming common in the 1990s. In 1941, an essay entitled 'Astrobiology' by

4

Laurence Lafleur (a philosopher at Brooklyn College, New York) described the word more narrowly than its modern incarnation as the consideration of life other than on Earth. Otto Struve, an astronomer at the University of California in Berkeley, also used the term in 1955 to describe the search for extraterrestrial life. A Russian astrophysicist, Gavriil Tikov (1875–1960), and a German astronomer, Joachim Herrmann, published books entitled *Astrobiology* in 1953 and 1974, respectively, covering popular ideas of extraterrestrial life.

The modern use of 'astrobiology' was introduced in 1995 by Wes Huntress, then at NASA's headquarters in Washington DC. At the time, NASA scientists argued that a study of life over scales from the microbial to the cosmic was essential for understanding life in the universe. Huntress liked the word *astrobiology* for this aspiration and the name stuck.

In fact, astrobiology was really a reinvention and expansion of *exobiology*, a field that goes back several decades. In 1960, Joshua Lederberg (1925–2008) coined the word exobiology for 'the evolution of life beyond our own planet'. Lederberg, a Nobel Prize winner for discoveries in bacterial genetics, argued that an essential part of space exploration should be searching for life. Then, from the 1960s onwards, NASA followed Lederberg's advice and financed exobiology research. But exobiology soon developed its critics. In 1964, George Gaylord Simpson, a Harvard biologist, quipped that 'this "science" has yet to demonstrate that its subject matter exists'.

The advantage of today's astrobiology is that it cannot be held to Simpson's charge because it includes the study of the origin and evolution of life on Earth at its core. Astrobiology also emphasizes the origin and evolution of planets as a context for life, and so embraces astronomers more firmly than exobiology.

Exobiology is not the only term similar to astrobiology. Since 1982, astronomers have officially used *bioastronomy* for the astronomical aspects of the search for extraterrestrial life, while earlier the word *cosmobiology* was favoured by J. Desmond Bernal (1901–71), an influential Irish-born British physical chemist. But neither of those terms has become widespread.

What is life?

Astrobiology raises the difficult question of how to define life. What is it exactly that we are looking for beyond Earth? A common approach is to list life's characteristics, which include reproduction, growth, energy utilization through metabolism, response to the environment, evolutionary adaptation, and the ordered structure of cells and anatomy. Unfortunately, this way of defining life is unsatisfactory for a couple of reasons. First, the list describes what life does rather than what life is. Second, most of these aspects of life are not unique. Life has structural order such as cells, but salt crystals are also ordered. Some of my friends have no children but they're alive, I think, as are mules that cannot reproduce. Growth and development apply to living entities, but also to spreading fires. All life metabolizes but so does my car. Life reacts to its environment, but a mercury thermometer also responds to its surroundings.

Alternatively, some scientists try to define life using *thermodynamics*, that is, heat and energy and their relationship to matter, by suggesting that the essence of life is the presence of stable structures, such as cells and genetic material, alongside *entropy* produced by metabolic waste and heat.

The term entropy requires some clarification. Desperate teachers searching for a quick and dirty explanation have often called it 'disorder'. Entropy is not disorder but an exact measure of energy dispersal amongst particles, be they atoms or molecules. Energy disperses spatially and so the energy of groups of particles that move together, which is said to be *coherent* energy, can dissipate. Thus, a bouncing ball comes to rest because its coherent energy of motion is converted into incoherent thermal motion of molecules and atoms through friction. In contrast, a stationary ball never spontaneously begins to bounce (as if it were alive) because even though sufficient thermal energy exists in the floor below, that energy is unavailable and dispersed in random jiggling of the atoms of the floor. The Second Law of Thermodynamics governs such phenomena and states that entropy in the universe never decreases. The entropy increase (or energy dispersal) conserves energy but ruins its quality. High-quality energy is not distributed but concentrated, such as in a barrel of oil, the nucleus of an atom, or in photons (particles of light) possessing high frequency and short wavelength. Such photons include ultraviolet and visible ones that cause sunburn and that power plant life, respectively. In physics, such high-quality energy has low entropy.

The physicist who most prominently linked entropy to life was the Nobel Laureate Erwin Schrödinger (1887–1961). In *What is Life?* (1944), Schrödinger commented that an organism 'tends to approach the dangerous state of maximum entropy, which is death. It can only keep aloof from it, i.e. alive, by continually drawing from its environment negative entropy... Indeed, in the case of higher animals we know the kind of orderliness they feed upon well enough, viz. the extremely well-ordered state of matter in more or less complicated organic compounds, which serve them as foodstuffs. After utilizing it they return it in a very much degraded form.' Regrettably, Schrödinger introduced the concept of 'negative entropy', which does not exist in science, to describe the ordered structure of food. Also, in the growth of some organisms, the increase in entropy primarily comes from heat generation rather than the degraded form of metabolic waste products compared to food. Linus Pauling (1901–94), who was arguably the greatest chemist of the 20th century, bluntly remarked that Schrödinger 'did not make any contribution whatever [to our understanding of life]... perhaps, by his discussion of "negative entropy" in relation to life, he made a negative contribution'.

Nonetheless a curious by-product of the ever-increasing entropy in the universe is that ordered, low-entropy structures, such as organisms, spring into existence. In fact, efficient production of entropy is best achieved by so-called *dissipative structures* involving a coherent *structure* of an immense number of molecules that *dissipates* energy. A simple example is a convection cell in boiling water. Warm water rises and is balanced by sinking water on its periphery. This circulating cell helps to disperse energy and so increases entropy more efficiently than if the cell were absent. All living organisms are complicated dissipative structures. However, attempts to define life with thermodynamics have so far failed to distinguish clearly between the living and non-living. For example, the writer Eric Schneider defines life as a 'far from equilibrium dissipative structure that maintains its local level of organization at the expense of producing environmental entropy'. A fire also fits this definition.

Pauling's criticism aside, Schrödinger argued correctly that organisms must run a sort of computer program, which is what we now call the *genome*. Indeed, life anywhere probably has to possess a genome. By a *genome*, we mean a heritable blueprint subject to small copying errors, which allows an organism to have evolved from an ancestor and provides a recipe for life's other characteristics such as metabolism. *Evolution*—the changes in populations over successive generations caused by selection of individuals' characteristics—is important because it is the only process that can explain the diversity of life and how the features of life that were listed previously were configured. In Darwin's natural selection mechanism, the genetic variation in populations of individuals means that some are better adapted for greater reproductive success than others. Natural selection favours genes that leave more descendants, so that lineages accumulate genetic adaptations.

Mindful of the centrality of evolution, astrobiologists often define life as 'a self-sustaining chemical system capable of Darwinian evolution'. Unfortunately, this definition is not helpful if we want to design an experiment to find life. Do we have to wait for evolution to happen for a positive detection? A better definition uses the past tense: 'life is a self-sustaining, genome-containing chemical system that *has* developed its characteristics through evolution'. So far, space-borne life detection experiments have not tried to measure a genetic make-up. For example, NASA's *Viking Lander* probes, which looked for life on Mars in the 1970s, were designed to recognize the Earth-like metabolism of microbes in the soil (Chapter 6).

The philosopher Carol Cleland and scientist Christopher Chyba have suggested that attempts to define life are like those of 17th-century scientists trying to define water. At that time, water was considered a colourless and odourless liquid that boils and freezes at certain temperatures. Without atomic theory, no one knew that water is a collection of molecules, each consisting of two atoms of hydrogen joined to an oxygen atom. By analogy, perhaps we lack the theory of living systems needed to define life.

Many of the problems in defining life boil down to the fact that we have only one example- life on Earth. All Earth-based organisms use nucleic acids for hereditary information, proteins to control biochemical reaction rates, and identical phosphorus-containing molecules to store energy. It's the same basic biochemistry in a bacterium or a blue whale. So it is difficult to distinguish which properties of life on Earth are unique and which are needed generally to qualify as 'life'. Astrobiology could help solve this conundrum if we found life beyond Earth.

The bare necessities of life

While there's no perfect definition of life, there are reasonable grounds to think that certain atoms common in terrestrial biochemistry are likely to be used by extraterrestrial organisms and might help us recognize life elsewhere. On Earth, the chief structural elements in biology are carbon, nitrogen, and hydrogen, while chemical interactions take place in liquid water. In astrobiology, there's wide agreement that life elsewhere is likely to be carbon based and that a planet with liquid water would, at least, favour 'life as we know it.' These deductions arise from realizing that life is constructed from a limited toolkit, the periodic table of chemical elements, which is the same throughout the universe.

In fact, carbon is the only element capable of forming long compounds of billions of atoms such as DNA (deoxyribonucleic acid). Consequently, only carbon-based extraterrestrial life seems able to have a genome of comparable complexity to terrestrial life. Carbon also has a variety of other properties that allow a unique chemistry of its compounds, sufficient to spawn the discipline of organic chemistry. Carbon's special properties include the ability to form single, double, and triple bonds with itself as well as bonds with many other elements. Carbon can also build three-dimensional complexity by forming hexagonal rings that join together.

Because life has to get started and propagate, it's probable that the main atoms of life are abundant ones. Carbon is fourth in cosmic abundance after hydrogen, helium, and oxygen. Indeed, astronomers have found that many non-biological organic molecules already exist in space. These freebies might serve as precursors to life getting started (see Chapter 3). For example, around 30 per cent by mass of the dust between the stars is organic material. So-called carbonaceous chondrite meteorites and interplanetary dust particles in our own Solar System contain up to 2 per cent and 35 per cent organic carbon by mass, respectively.

Because silicon has chemical properties similar to carbon, it is sometimes asserted that silicon might allow an alternative extraterrestrial biochemistry to carbon-based molecules, despite being about ten times less cosmically abundant than carbon. But in water, at least, silicon compounds tend to be unstable and silicon easily gets locked into solid silicon oxides. Carbon dioxide is a gas at common planetary temperatures and dissolves in water to concentrations sufficient for organisms to use carbon dioxide as a carbon source. Silicon dioxide, in contrast, is an insoluble solid, such as quartz. Silicon's bonds with oxygen and hydrogen are strong, whereas carbon–oxygen and carbon–hydrogen bonds are similar in strength to the carbon–carbon bond, which allows carbon-based compounds to undergo reactions of exchange and modification. Silicon–hydrogen bonds also tend to be easily attacked in water. The stability of silicon-based molecules requires low temperatures to slow down reactions that would otherwise destroy them. Appropriately cold solvents include oceans of liquid nitrogen on icy planets far from their stars. At present, such silicon-based life remains purely speculative.

However, a stable medium is necessary for biochemical processes such as metabolism or genetic replication; on Earth this medium is liquid water. For extraterrestrial life, the medium could be another liquid or a dense gas as long as it doesn't easily become a solid in the prevailing environment. Nonetheless, water (H_2O) has some unique properties. Unlike its smelly twin, hydrogen sulphide (H_2S), which condenses to a nasty liquid only at $-61°C$, water turns to liquid below $100°C$ at normal pressures. Liquid water's stability occurs because the oxygen atom in water molecules is slightly negatively charged and allows a relatively strong 'hydrogen bond' to the slightly positive hydrogen atoms of other water molecules. Sulphur provides weaker hydrogen bonds between hydrogen sulphide molecules. Water also forms stronger hydrogen bonds to other water molecules than to molecules of oily substances. As a consequence, oils separate from water, which allows cell membranes to form and provide homes for genes and metabolic processes.

Another unusual property of water is that ice is less dense than liquid water. When water freezes, the molecules align into ring-like structures containing open holes on the atomic scale. If ice were denser than liquid, the cold bottom of lakes and seas would collect ice, which would be insulated and remain frozen. Seas would freeze from the bottom up and become uninhabitable. This would arise because sunlight would be reflected back in the areas where the ice reached the surface, causing cooling and more ice to accumulate. Gradual freezing might be the unfortunate fate of seas of other liquids such as ammonia. Ammonia is a liquid from about −78°C to about −33°C at a pressure of one atmosphere. But any seas of ammonia would tend to solidify from the bottom up, unlike seas of water that remain liquid even if cold temperatures cause an ice cover.

On Earth, we find microbes wherever there's liquid water (excluding sterilized apparatus), so 'life as we know it' is as much 'water based' as 'carbon based.' Consequently, for Solar System exploration, the detection of liquid water or its past presence provides an objective for planetary probes, such as those visiting Mars. Nonetheless, it is possible to conceive of organic solvents as alternatives to water, which is potentially important for Titan, Saturn's largest moon (Chapter 6).

Another observation from terrestrial life is that just six non-metallic elements—carbon, hydrogen, nitrogen, oxygen, phosphorus, and sulphur—make up 99 per cent of living material by mass. These elements are often abbreviated as 'CHNOPS', but I prefer 'SPONCH', which is easier to say. We find H and O in water, which makes up most living tissue, and C, H, and O in the nucleic acids of genetic material and in carbohydrates. C, H, N, and S exist in proteins, while P is essential for nucleic acids and energy-storage molecules. Consequently, detecting the SPONCH elements in chemical forms that life could use is another practical goal for planetary space probes searching for Earth-like life.

The significance of life elsewhere

You might wonder whether the discovery of simple, microbial-like extraterrestrials would really matter. But if we could find a single instance of life that originated elsewhere, it would prove that life is not a miracle confined to Earth. We wouldn't be alone. Even the simplest microbes native to Mars or to Europa's oceans would change the balance of probabilities that life exists elsewhere in the galaxy for they would demonstrate that life can originate twice within one solar system. At the moment, we have no convincing evidence of life beyond Earth. However, in Chapter 6, I will argue that at least nine other bodies in our Solar System might be habitable today, if we keep an open mind. Solar System astrobiology is far from settled.

A second significant factor in finding life harks back to the difficulties of defining life. The planetary scientist Carl Sagan (1934–96) commented in his book *Cosmic Connection* (1973) that 'the science that has by far the most to gain from planetary exploration is biology'. An examination of extraterrestrial life would be of profound significance not only in identifying those elusive characteristics common to all life but also in shedding light on how life originated, which remains unsolved.

Chapter 2
From stardust to planets, the abodes for life

> To make an apple pie from scratch, you must first invent
> the universe.
>
> Carl Sagan (1980)

When the universe began, temperatures everywhere were far too
hot for atoms to be stable, let alone join up into complex biological
molecules. Life exists because following a *Big Bang* 13.8 billion
years ago, a hot, dense cosmos expanded and cooled. As it did so,
atoms, galaxies, stars, planets, and life arose. Here, we examine
how this process produced an abode for life—the Earth.

We start with the structure of the present universe, which
provides the clues about its history. Imagine a journey that goes
to the edge of the observable universe. At the speed of light—
300,000 km per second—it would take only 1.3 seconds to get to
the Moon at its distance of 384,000 km. A diagram of the Earth
represented as a spot of 2.7 mm diameter and the Moon looks as
follows:

●   •
Earth   Moon

This scale is fairly easy to grasp. But it challenges our imagination
when we realize that the Sun on the same scale would be 30 cm in

diameter and we would have to place it just over 30 m away from this book. The nearest star, Proxima Centauri, which is small compared to the Sun, would be about 4 cm in diameter. To keep to scale, we would have to place it 8,650 km away, which is roughly the flight path from San Francisco to London.

Continuing our voyage at the speed of light, it would take 8.3 minutes to fly from the Earth to the Sun and a little over four hours to then travel to the average orbital distance of Neptune, the outermost of the eight planets. After 4.2 years, we would reach Proxima Centauri. With the huge distances involved, we define the distance travelled in a year at the speed of light, some 9,500 billion km, as a *light year*, so that Proxima Centauri is 4.2 light years away.

The Sun and Proxima Centauri are two of about 300 billion stars in the Milky Way galaxy, which is a disc some 100,000 light years across with stars concentrated in spiral arms. Galaxies contain millions to trillions of stars, so the Milky Way is moderately large. The Solar System sits in the Orion Arm, two-thirds out from the Galactic Centre. This arm appears unremarkable compared to some others that are more richly populated with stars. But it's possible that the Solar System's location in the galactic sticks was actually vital for terrestrial life. The Earth may have avoided certain catastrophes, such as proximity to exploding stars. If so, there may be a particular region in galaxies favourable for life, called the 'Galactic Habitable Zone', discussed in Chapter 7.

On a larger scale, there are more than 100 billion galaxies in the observable universe arranged into groups and superclusters. Within a diameter of about 10 million light years centred between the Milky Way and our nearest spiral galaxy, the Andromeda galaxy some 2.5 million light years away, there are roughly fifty galaxies, forming the *Local Group*. In turn, this group is one of a hundred or so within a sphere of 110 million light years' diameter, comprising the *Virgo Supercluster*. Maps that cover billions of light years show filaments of tiny scattered points where each point is a galaxy. In three dimensions, galactic filaments, which are the largest structures known to humankind, join up into a web separated by vast voids. The whole fantastic structure looks as if it were spun by a crazy intergalactic spider.

How big is the observable universe? If space had not expanded, the farthest distance would be 13.8 billion light years, which is that traversed by a photon—a particle of light—since the Big Bang happened 13.8 billion years ago. But space has expanded. So the actual size of the observable universe is now about 47 billion light years across. Such vastness is a consideration for astrobiology because surveys of planets around stars just within the Milky Way suggest that each star hosts at least one planet on average. Some fraction of planets ought be habitable, perhaps at least 1 per cent, so the number of potential abodes for life might exceed a trillion billion.

The structure of the universe traces back to the Big Bang. In the 1920s, telescopic observations by the American astronomer Edwin Hubble showed that galaxies are moving away from each other on the large scale as space expands between them. Thus, going back far in time, everything must have been scrunched up and very hot. This consideration led to the recognition of strong evidence for the Big Bang. Its afterglow, the *Cosmic Microwave Background*, permeates the entire universe. If the Big Bang is true, physics dictates that before the early universe cooled, it must have been an opaque fireball made of electrons and protons (the electrically negative and positive elementary particles that make up atoms), photons (particles of light), and some small groups of fused protons. Some 380,000 years after the Big Bang, it became cool enough that electrons were able to join with protons, or groups of protons, to form the two smallest atoms, hydrogen and helium. At that point, the universe became transparent to light because previously photons had been scattered by the free-floating electrons. This prior situation was analogous to the way that light bounces around off tiny droplets in a fog so that you can't see through. Since losing its opaqueness, the universe has expanded by about a factor of 1,000, and the wavelength of the relic photons from the Big Bang has been stretched the same amount, changing the photons from red light to microwave. Amazingly, when you tune an old analogue television or radio between channels, the leftover radiation from the Big Bang contributes to the static hiss, albeit at a level of about 1 per cent or less. In fact, in 1964, this noise in a large radio horn was how the microwave background was discovered. Pigeon droppings in the horn were blamed initially; but after cleaning up and shooting the unfortunate pigeons, the real culprit was identified as the beginning of the universe.

Galaxies appeared a few hundred million years after the Big Bang. Some places had very slightly more material than average and so had higher gravity. Clumping produced galaxies, and within the galaxies, on a smaller scale, gas clouds collapsed under their own gravity. The interior of each shrinking cloud heated up as gas particles collided, eventually making a hot, glowing ball of gas—a star.

Part of the 'astro' in astrobiology comes from the fact that all of the atoms used by life except for hydrogen were created inside stars. The first stars would have been made only of the elements synthesized in the Big Bang: hydrogen, comprising three-quarters of the mass, and helium, which made up the rest except for a trace of lithium. The oxygen in water, the nitrogen in proteins, or the carbon in every organic molecule—none of these elements was present at first. But eventually stars made them.

To understand how stars make elements, consider how the Sun shines. Inside the Sun, immense heat strips each atom down into its constituents: a positively charged nucleus and negatively charged electrons. The temperature at the centre of the Sun, some 16 million degrees Celsius, is enough to fuse the nuclei of four hydrogen atoms into a helium nucleus, which is a nuclear reaction that releases photons. Each photon then endures a one-million-year journey from the interior of the Sun to space. It takes so long because each photon is continually absorbed and emitted as it encounters material. The photon also loses energy. It starts out as a high-energy gamma ray and on average turns into a lower-energy photon of visible light by the time it escapes from the Sun. In the 1950s, physicists reproduced the kind of nuclear reactions that occur inside the Sun with hydrogen bombs. Cores of stars like our Sun are akin to hydrogen bombs that, in effect, can't explode because they are contained by the weight of material above them.

The Sun will not fuse hydrogen in its core forever and the repercussions will destroy life on Earth (partly answering astrobiology's question of 'What's the future of life?'). In most stellar cores, the accumulation of helium 'ash' causes temperatures to drop too low to support further hydrogen fusion. At this point, the star shrinks under its own weight, which causes the temperature to rise until it ignites hydrogen fusion in a shell surrounding the core. The energy release causes outer layers of the star to expand, cool, and redden. This is how a *red giant* forms, such as Aldebaran, the brightest star in the constellation of Taurus, the Bull. The Sun will eventually become a red giant and swell two-hundredfold by 7.5 billion years' time, probably engulfing the Earth. The weight of further helium ash eventually squeezes a red giant's core to a temperature of 100–200 million degrees Celsius, which is enough to fuse helium nuclei and make carbon and oxygen. In turn, a helium-burning shell can eventually surround a core of carbon and oxygen 'ash'. Stars with four to eight times the mass of the Sun even end up fusing the carbon and oxygen into heavier elements, including neon and magnesium.

Generally, the death throes of a Sun-like star involve shedding outer layers into space. These shells of glowing gas are called *planetary nebulae* because they look somewhat like planets through low-magnification telescopes, but they have nothing to do with planets. The remnants of the star cool down into a *white dwarf* with a radius comparable to that of the Earth but an enormous density. In theory, a white dwarf stops shining after tens to hundreds of billions of years, producing a *black dwarf*. But the universe is not yet that old.

Stars larger than about eight times the mass of the Sun eventually explode as *supernovae*. The Sun is about half way through its ten-billion-year phase of hydrogen fusion in its core, but these massive stars spend less than 60 million years in the same phase before becoming *red supergiants*, of which Betelgeuse, in the constellation of Orion, is an example. In such stars, nuclear fusion in shells around the core produces the elements neon, magnesium, silicon, and iron. Iron is generally the heaviest element made, although some heavier elements are also produced when free neutrons are added to existing nuclei. (Neutrons are particles that have no electrical charge and are commonly found in atomic nuclei.) When the fuel in such a star runs out, the core is so compressed that negatively charged electrons amalgamate with the positively charged protons of the iron nuclei. The electrical charges cancel out and uncharged neutron particles are created. As a result, the stellar core shrivels to about 12 km diameter, forming something like a gigantic atomic nucleus made only of neutrons. Because of the shrinkage, the rest of the star collapses onto the dense neutron core, and a violent rebound creates the incredibly bright supernova.

Elements heavier than iron are generated and distributed by supernovae. A few seconds after a supernova, the outer layers of the star are heated to an incredible 10 billion degrees Celsius while the break-up of nuclei from deeper layers supplies abundant neutrons to fuel reactions that make heavy elements. This cosmic alchemy creates the heavy, precious elements such as gold, silver and platinum. Importantly, the material blown out in a supernova

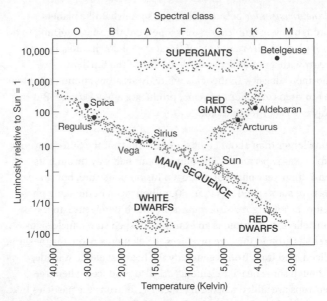

Spectral class

1. **The Hertzsprung–Russell diagram for the Sun and nearby stars. Temperatures are in Kelvin, which is 273 plus the temperature in degrees Celsius; the Kelvin scale is defined so that 0 K or 'absolute zero' is where all molecular movement stops**

forms the basis of new generations of stars and abodes for life, their planets. In the case of stars reaching tens of solar masses, although a supernova still occurs, the collapse at the centre produces a *black hole*—an object so massive that nothing escapes its gravity, including light.

The phase when a star converts hydrogen to helium in its core is the *main sequence* lifetime, which is generally the interval we consider optimal for life to thrive on planets around the star. The 'main sequence' refers to a diagonal swathe from upper left to lower right on the most famous graph in astronomy, the *Hertzsprung–Russell (H–R) diagram* (Fig. 1), named after its two originators. The graph plots a star's luminosity versus its surface temperature. By a 'surface', astronomers don't mean a hard surface but the level in a star's atmosphere where most light emerges, which is as deep as we can see.

Oddly, the temperature axis runs backwards from high to low in the H–R diagram. The purpose is to match the colour coding of stars from hot blue-white stars to cooler red ones with the letters O, B, A, F, G, K, and M. The letter gives a star's *spectral class*. Generations of astronomy students have remembered spectral types with the mnemonic 'Oh Be A Fine Girl/Guy Kiss Me!' (the letters don't stand for anything and have origins in 19th-century astronomy, which need not concern us).

On the main sequence, the most massive stars plot at the upper left and the lightest at the lower right. Whatever its mass, a main sequence star is called a dwarf—such as the Sun, a G-type dwarf. The lifetime on the main sequence can be over 50 billion years for cool red dwarfs.

The importance to astrobiology of the way stars 'live' and 'die' is wide ranging. Our Sun is a middle-aged, main sequence star with stable sunlight that fuels most life on Earth through photosynthesis. As mentioned at the outset, the atoms of life were generated in red giants and supergiants. Also, oxygen, silicon, magnesium, and iron are made from integral numbers of helium nuclei and so are particularly abundant products of nuclear fusion, which is significant because these atoms are the ones that make rocks. Rocky planets, like the one we inhabit, are a natural consequence of the physics of starlight. We also know that the Sun is at least a second-generation star because we have supernova elements on Earth such as gold. But since the Sun is only 4.6 billion years old in a 13.2 billion-year-old Milky Way, many stars came and went before life arose on Earth. Did earlier stars support planets and life, or even intelligent life, and what happened to them? This leads us to the question of how our own planet formed.

Getting a place to live: where planets come from

Ideas for the origin of the Solar System have a long heritage. In 1755, Immanuel Kant suggested that the Solar System coalesced out of a diffuse cloud in space. Later, in 1796, the mathematician Pierre Simon, the Marquis de Laplace, elaborated the notion. The basic Kant–Laplace concept is known as the 'nebular hypothesis,' after *nebula*, the Greek word for cloud.

The hypothesis starts with the idea that some part of the cloud is slightly denser than others and attracts material by gravity. It is likely that the cloud also has some slight initial rotation, so that the shrinking cloud spins faster, like an ice skater drawing in her arms. The random motion of the gas and dust will oppose material attracted along the axis of rotation but matter converging in the plane of rotation will also be resisted by the spin. As a result, the cloud flattens into a disc. The Sun forms in the centre, while planets coalesce in the plane of the disc from sparse material. This is consistent with the mass of the planets being only 0.1 per cent of the Sun's mass. Because they form from a disc, planets will all be in the same plane and they will all orbit the Sun in the same direction, as we observe.

In recent decades, it was thought that evidence from isotopes suggested that a shock wave from a nearby supernova triggered the collapse of the nebula. Isotopes are atoms that contain the same number of protons in their nucleus but a different number of neutrons. The Greek root, *isos topos*, means 'equal place', which refers to the same location in the periodic table of elements. Sometimes an isotope has a nucleus that is too big to be stable, and breaks apart by radioactive decay. An unstable aluminium-26 atom (which has a nucleus of 26 particles = 13 protons + 13 neutrons) decays into stable magnesium-26 (containing 12 protons + 14 neutrons). In any sample of aluminium-26 atoms, the time it takes for half of them to change into magnesium-26, the *half-life*, is 700,000 years. The presence of magnesium-26 in some meteorites, which must have been produced relatively quickly from aluminium-26, was thought to suggest a nearby supernova when the Solar System formed because massive stars make aluminium-26 and their supernovae distribute it.

However, in 2012, new measurements in meteorites showed that the levels of iron-60, which is an isotope only formed in supernovae, are too low for a nearby supernova. To explain abundant aluminum-26, a neighbouring massive star (perhaps more than twenty times the solar mass) of a type known as a Wolf–Rayet star may have shed its outer layers and spread aluminum-26 into the solar nebula. Aluminum-26 was a dominant heat source in the solar nebula for the first few million years. By melting ice in the earliest rocky material, the aluminum-26 caused water to go into hydrated minerals where it was safe, unlike ice that evaporates. If the aluminum 26 had not been present, the Earth might not have gained water-rich minerals and it might have neither oceans nor life.

The nebular hypothesis explains the broad distribution of different types of planet. The planets in the inner Solar System (Mercury, Venus, Earth, and Mars) are relatively small and rocky, while those in the outer Solar System (Jupiter, Saturn, Uranus, and Neptune) are giants. When the disc was forming, the matter drawn inwards gained energy of motion and the centre of the disc, where the Sun formed, became extremely hot. This created a temperature gradient from the hot centre of the disc to a cold exterior. Outside the Sun, pressures were low, which meant that substances existed either as solids or gases but not liquids. In the inner disc, it was far too hot for gases such as water vapour to form ice, but metal and rock vapour could condense nearly anywhere in the disc. Consequently, planets that formed in the inner region, such as the Earth, ended up with iron-rich cores surrounded by rocky mantles. In contrast, water condensed into ice from just inside the orbit of Jupiter ('the ice line') and farther out, providing more material to make larger planets. Methane could also condense as an ice from the orbit of Neptune outwards.

The giant planets are considered to have formed before the rocky ones. Jupiter and Saturn probably formed when rocky cores reached a size of about ten or so Earth masses that had sufficient gravity to attract more and more hydrogen and helium gas directly from the disc until all that was available in their orbit was sucked up. Because they are huge balls of mostly gas, we call these planets *gas giants*. This process happened within about 10 million years after the formation of the Sun. Uranus and Neptune are smaller, and drew in a greater proportion of icy solids, so we call these planets *ice giants*. Unlike the rapid growth of Jupiter, the formation of the inner rocky planets was spread over 100–200 million years. Planetesimals, which are 'pieces of planet', coalesced to make larger, rocky objects called planetary embryos with a size between that of the Moon and Mars. In turn, several planetary embryos merged into Venus and Earth and fewer into Mercury and Mars.

Although the inner planets accumulated material in their locality, gravitational nudges from the giant planets, particularly Jupiter, would have sent planetesimals careering into the inner Solar System. Scattered hydrated asteroids that originated from beyond the orbit of Mars were likely responsible for bringing water to the Earth that eventually turned into our oceans and lakes. We drink asteroid water. Computer simulations show that Jupiter ejected more water-rich material than it scattered inward, so that if Jupiter had had a less circular orbit and ejected even more water-rich material, Earth might have ended up without oceans and life.

We also know that planets do not necessarily stay put in their orbits. So-called *hot Jupiters* are exoplanets similar in mass to Jupiter that orbit at least twice as close to their host stars as the Earth orbits the Sun. Such exoplanets cannot have formed where they currently reside because it would have been too hot. It turns out that planets migrate because of large tails of gas and dust in the nebula that accompany their formation as well as the gravitational influence from other bodies. So the traditional nebula hypothesis is nuanced by *planetary migration* that could destroy or favour planetary habitability in extrasolar systems, depending on the details. In the next chapter, I'll discuss the idea that even the giant planets of our own Solar System may have migrated somewhat.

Nonetheless, the essential idea of the nebular hypothesis—that the planets formed from a disc—was confirmed in the 1980s when discs of debris around young stars were seen through telescopes. In fact, there is still some leftover debris in our own Solar System. Comets are icy bodies and asteroids are rocky rubble that were never assimilated. Occasionally, small chunks that have been knocked off asteroids through collisions end up falling onto the Earth's surface as meteorites. Consequently, meteorites provide us with key data about the early Solar System.

The age of the Earth and the Moon

Perhaps the most profound information gleaned from meteorites is the age of the Earth and Solar System. Ever since the 18th century, the vast layers of sedimentary rock seen by geologists had led them to suspect that the Earth, and hence the Solar System, must be of great age, but proof was lacking.

The first person to attempt to measure the age of the Earth was an English geologist, Arthur Holmes, who had the idea of examining lead isotopes formed from radioactive uranium. Uranium-238 decays into a cascade of further unstable isotopes of other elements until reaching a stable lead isotope, lead-206. The half-life for uranium-238 to change into lead-206 is 4.47 billion years. Another radioactive isotope, uranium-235, decays into lead-207 with a half-life of 704 million years. So, by measuring the amounts of lead-207 and lead-206 in different mineral grains of a rock, you can determine the age of the rock, because although there may have been different amounts of uranium in each grain, the fixed decay rates add the same ratio of lead-206 to lead-207 in a given time. In 1947, Holmes applied his method to a piece of lead ore from Greenland, and estimated that the Earth formed by 3.4 Ga (where Ga means 'Giga anna', equivalent to 1,000 million years or 'billions of years ago').

Holmes had two problems. First, he could never be sure that even the oldest rock he could find was as old as the Earth itself. Second, for a precise age, Holmes needed to know the small, original ratio of lead isotopes present when the Earth formed, so-called 'primeval lead', before subsequent uranium decay added further lead atoms. Clair Patterson, an American geochemist, realized that the first problem could be avoided by looking at meteorites because these are leftover building materials that formed around the same time as the Earth. He also recognized that certain types of meteorite, the iron meteorites, contain negligible uranium and so their ratio of lead isotopes provides a measure of primeval lead. With this approach, in 1953, Patterson accurately aged the Earth at 4.5 Ga. He was so excited that he feared a heart attack, and his mother had to take him to hospital.

Since then, improved techniques using radioactive isotopes have given us a timeline for the events surrounding the Earth's formation. The very oldest grains in meteorites suggest that the Solar System formed at 4.57 Ga. The Earth formed slightly later at 4.54 Ga. Then, around 4.5 Ga, the Earth was apparently hit by a Mars-sized object, which is named *Theia* after the Greek goddess who gave birth to the Moon goddess Selene. According to this *giant impact hypothesis*, debris that was blasted out from the impact went into orbit around the Earth and coalesced to form the Moon.

The Moon is important for astrobiology because its gravity stabilizes the tilt of Earth's axis to the plane of its orbit, which helps the Earth maintain a relatively steady climate. Currently, the Earth's axis is tilted 23.5 degrees, but if it were able to vary widely, great climatic swings would occur. For example, at 90 degrees tilt, the Earth would be tipped on its side and ice would form seasonally at the equator. Computer simulations show that if we had no Moon, the Earth's axial tilt would vary chaotically over periods of millions of years and have a large range, potentially from 0 to more than 50 degrees. Microbial life could probably withstand large climate swings. But advanced animal life and civilizations such as ours would be challenged.

In following our astronomical trail, we have now arrived at the point of the creation of an abode for life, our own planet. There were contingencies in whether the water necessary for life was delivered to the Earth and controls on where an Earth-like planet might occur. Then, after Earth formed, how and when did life arise?

Chapter 3
Origins of life and environment

The early Earth

So little is known about the very earliest aeon of Earth history that it doesn't even have an official name. Informally, it is called the Hadean, which started around 4.5 Ga when the Moon formed, and ended at a date that hasn't been agreed upon but is usually taken as either 3.8 or 4.0 Ga. It is probable that life originated in the Hadean, but so far we haven't found evidence because there are no sedimentary rocks from this time. Such rocks consist of layers of sediment grains laid down in water or from airfall, and so best preserve traces of biology or the environment. Consequently, our ideas of what happened during the Hadean have to be constructed from sparse data aided by the theoretical constraints.

Theory suggests that heat from the giant impact that formed the Moon would have turned rock into gas. The atmosphere of vaporized rock would have lasted for a few thousand years and then condensed and rained down on a molten magma surface, which eventually solidified into a crust. Subsequently, the atmosphere would have consisted mainly of extremely dense steam for a few million years before it condensed to form oceans.

When did continents begin to form? We get some clues from tiny mineral grains less than 0.5 mm across that are left behind from the first half billion years or so of Earth history. These are *zircons*, crystals of zirconium silicate with chemical formula $ZrSiO_4$, where Zr is zirconium, Si is silicon, and O is oxygen. Zircons are so tough that they remain even after the rock in which they were once hosted has eroded away. Some zircons as old as 4.4 to 4.0 Ga have been found in fossilized gravel in the Jack Hills, which are about a thousand kilometres north of Perth in western Australia. The zircons contain flecks of quartz, which is a crystalline form of silica, SiO_2. The quartz may have been derived from granites, which is the silica-rich type of igneous rock that makes up much of the continents. In this way, zircons suggest that Earth's continental crust existed as long ago as 4.3 Ga. Isotopes support this inference. Some ancient zircons are enriched in stable oxygen-18 atoms relative to stable oxygen-16 ones. Such enrichment occurs when surface waters make clays and the mud is then buried and melts underground, passing the isotopic signature to igneous rocks.

In the Hadean, the Earth was probably hit by a few huge pieces of debris left over from Solar System formation, but none as big as the Moon-forming impactor. The energy of a small number of very large impacts could have vaporized the entire ocean or its upper few hundred metres. If so, life would have had to restart or it might have been trimmed back to only those microbes sheltered underground that were able to survive the heat. In fact, partially sterilizing impacts may explain the nature of the last common ancestor for all life on Earth. Genetics traces the common ancestor to a thermophile (see Chapter 5)—a microbe that lives in hot environments. Essentially, DNA analysis implies that your 'great-great-great-…grandmother' was a thermophile, if you insert enough 'greats'. This may be because thermophiles were the only survivors of huge impacts.

The Earth today is not entirely free from potentially catastrophic impacts. For example, Chiron is a comet-like object about 230 km across in the outer Solar System that crosses the orbit of

Saturn. One day within the next 10 million years or so, a nudge from Saturn's gravity will fling Chiron towards the Sun or away from it. In the former case, Chiron's chances of impacting the Earth would be less than one in a million. But if Chiron did hit, the heat would turn the upper few hundred metres of the ocean into steam, and land would be sterilized down to about fifty metres' depth. Thermophiles might become the ancestors of subsequent life in a sort of evolutionary déjà vu, if the impact explanation for a thermophile ancestor is correct.

The last hurrah of the big Hadean impacts is known as the *Late Heavy Bombardment*, which happened about 4 to 3.8 Ga. Craters on the Moon bear witness to this massive bombardment. Rocks brought back by Apollo astronauts have been dated using radioisotopes and indicate that many craters were produced within the same 200-million-year interval.

The leading hypothesis for explaining the Late Heavy Bombardment was devised by astronomers in Nice, France, and is therefore called the *Nice model*. It relies on the astonishing idea that the orbits of Jupiter and the other giant planets shifted at the end of the Hadean. Calculations suggest that after the Solar System formed, mutual gravitational effects caused Saturn and Jupiter to reach a state called a 'resonance' in which Saturn orbited the Sun once for every two Jupiter orbits. Regular alignment created periodic gravitational prodding that made the orbits of Saturn and Jupiter less circular. In turn, the orbits of Neptune and Uranus were perturbed and moved outwards, also becoming more elliptical. Neptune could have even begun inside the orbit of Uranus and then overtaken it in a migration outward, and both planets, particularly the farther one, would have scattered small icy bodies, some towards the inner Solar System. Meanwhile, the gravity and movement of Jupiter would fling some asteroids into the inner Solar System and push others away. During this time, the Earth must have suffered even more impacts than the Moon because of Earth's greater size and gravity.

Eventually, after wreaking havoc, the orbits of the giant planets would have settled down.

The origin of life

Quite how life arose is unknown. It may have originated on Earth or it was carried here by space dust or meteorites. The latter idea is called *panspermia*, and doesn't solve how life originated, but pushes the problem elsewhere. Also, there may be difficulties with survival over long transit times if life came from around other stars. For these reasons, we'll concentrate on a terrestrial origin of life.

There is wide agreement that the origin of life would have been preceded by a period of chemical evolution, or *prebiotic chemistry*, during which more complex organic molecules were produced from simpler ones. The idea goes back to the 19th century. In 1871, Charles Darwin imagined that such chemistry might have occurred in a 'warm little pond', according to a letter to the botanist Joseph Hooker:

> It is often said that all the conditions for the first production of a living organism are now present, which could ever have been present.—But if (and oh what a big if) we could conceive in some warm little pond with all sorts of ammonia and phosphoric salts,—light, heat, electricity etc. present, that a protein compound was chemically formed, ready to undergo still more complex changes, at the present day such matter would be instantly devoured, or absorbed, which would not have been the case before living creatures were formed.

Thus, Darwin concluded that life is unlikely to originate today because organisms are continually eating the chemical compounds that are needed. On the other hand, before life existed, chemical conditions for the origin of life would have been prevalent.

In the 1920s, the Russian biochemist Alexander Oparin and the British biologist J. B. S. Haldane both recognized that the environment under which life arose would have lacked oxygen. Earth's oxygen-rich atmosphere is a product of photosynthesis by plants, algae, and bacteria. An oxygen-free atmosphere would have been better suited to prebiotic chemistry because oxygen converts organic matter into carbon dioxide, which prevents the build-up of complex molecules. Indeed, a contemporary geologist, Alexander MacGregor, reported in 1927 that he had found sedimentary rocks dating from within the Archaean Aeon (3.8–2.5 Ga) that showed that the ancient atmosphere lacked oxygen. In particular, MacGregor observed that iron minerals in what is now Zimbabwe were unoxidized, unlike the rust-coloured iron oxides in modern sediments that are produced when atmospheric oxygen reacts with iron-containing minerals. Today, MacGregor's deduction is supported by many other data (Chapter 4).

Oparin and Haldane also proposed that gases in Earth's early atmosphere would have been converted by ultraviolet sunlight or lightning into organic molecules. In the 1950s, such ideas were tested when Stanley Miller, a student of the Nobel Prize-winning chemist Harold Urey, devised an experiment at the University of Chicago. A blend of ammonia, methane, hydrogen, and water vapour was put in a flask and electric sparks passed through the gas to simulate lightning. Miller found that yellowish water condensed out at the bottom of the flask. This dark residue contained organic molecules, including various amino acids—the building blocks of proteins. At the time, the Miller–Urey experiment was trumpeted in the media as virtually solving the problem of the origin of life. Life was said to start in a 'primordial soup' generated by atmospheric chemistry. In 1953, when Miller wrote up his results, a common view (dating back to Oparin and Haldane) was that genetic material was protein, but a month before Miller's paper came out, DNA was identified as the real basis of heredity. Subsequently, experiments in the Miller–Urey genre have also produced molecules containing hexagonal rings of carbon, nitrogen, and hydrogen atoms, which are the kind of rings found in DNA.

However, geochemists have raised doubts about the Miller–Urey experiment because they argue that the Earth's early atmosphere was probably not as hydrogen rich as Miller assumed. Volcanoes supply the Earth's atmosphere with gases over long timescales but most volcanic gas is steam, i.e. water vapour (H_2O), with an average of less than 1 per cent hydrogen (H_2). Similarly, volcanic carbon comes out as carbon dioxide (CO_2) rather than the hydrogenated form of methane (CH_4), while nitrogen is emitted as dinitrogen (N_2) instead of ammonia (NH_3).

It is thought that the atmosphere in the Hadean was mostly maintained from gases released through volcanism, so the air should have consisted predominantly of CO_2 and N_2. Of course, it depends on how far back you go. When the Earth was forming, the vaporization of large impactors would have produced hydrogen-rich atmospheres if impactors were sufficiently rich in iron to provide the right kind of chemistry to stabilize hydrogen in the impact explosion. Early on, most of the atmosphere was steam and oceans had not yet condensed. After oceans formed, atmospheres produced by impact vaporization would have been ephemeral. Because of uncertainties about the composition of the Hadean atmosphere, astrobiologists have proposed sources of organic carbon other than that presumed by Miller and Urey.

One alternative is that organic carbon came from space. Some carbon-rich meteorites contain amino acids, alcohols, and other organic compounds that perhaps could have seeded Earth with the materials needed for prebiotic chemistry. For example, the Murchison meteorite, which fell on Murchison, Australia, in 1969, contains over 14,000 different molecules. Also, micrometeorite particles that are 0.02–0.4 mm in size currently bring about 30 million kg of organic carbon to the Earth per year. Organic carbon can fall intact to the Earth's surface because tiny particles don't necessarily burn up in the atmosphere. Furthermore, the interiors of comets are rich in organic material and may have played a similar role to meteorites in getting life started if the organic material survived cometary impacts on the Earth.

Another possible source of organic carbon is deep-sea hydrothermal vents. A hydrothermal vent is a spot where hot water emerges from the seafloor. New seafloor is created at mid-ocean ridges, where the Earth's tectonic plates separate and magma rises up from the mantle, filling the void. Seawater flows down through cracks near the ridges, is heated by the hot magma, and rises, picking up substances produced by chemical reactions between the water and rocks, such as hydrogen, hydrogen sulphide, and dissolved iron. When the hot acidic water hits cold (2°C) ocean water, it produces a plume of precipitated particles containing dark-coloured pyrite, an iron sulphide mineral (FeS_2, where Fe is iron and S is sulphur). That is why these hydrothermal vents are called 'black smokers'. One hypothesis suggests that life originated on the surface of such pyrite minerals. However, black smokers are very hot, around 350°C, so most researchers who favour a deep-sea origin of life have focused on vents that are cooler, around 90°C, alkaline rather than acidic, and located further from the mid-ocean ridges.

Beneath alkaline vents, hydrogen released in reactions between water and rock can combine with carbon dioxide to make methane and bigger organic molecules. In the Hadean, tiny pores in the mineral structures that grew up from vents could, in principle, have contained and concentrated simple molecules to allow them to react and make more complex prebiotic molecules.

Also, in alkaline vents, because the fluid emerging is more alkaline than seawater, there is a natural gradient in pH that is similar to that in cells. If life originated in such an environment, it might explain why energy production is intimately tied to pH gradients.

Life generates energy from microscopic electrical motors that are embedded in cell membranes and run off electrical currents driven by pH gradients across the membranes. It is impossible for words to do justice to these amazing molecular machines, so I suggest that the reader search the Internet for 'ATP synthase animations'. Essentially, metabolic energy is used to create a different pH on each side of a cell membrane with the outside usually more acidic than the inside. This differential is a *proton gradient*, because pH is an inverse measure of the concentration of positively charged hydrogen ions (protons) in solution—the greater the concentration, the lower (and more acidic) the pH. The proton gradient is essentially a battery. When discharged, electrical current flows through the molecular turbine in the cell membrane and generates molecules that store energy. These molecules of adenosine triphosphate or 'ATP' are the energy carriers in all cells. For example, a human typically contains 250g of ATP and uses up a body weight's worth of ATP every day as ATP is made used, and remade.

The above process of generating ATP is called *chemiosmosis*, which contains the word 'osmosis', meaning water moving across a membrane, because the proton movement is analogous. Chemiosmosis is so weird compared with the familiar idea of generating energy from chemical reactions that when it was proposed in 1961 by British biochemist Peter Mitchell most biochemists were hostile. Eventually, Mitchell was vindicated with a Nobel Prize in 1978. Given that chemiosmosis is universal to terrestrial life and presumably related to its origin, it is interesting to wonder whether ion gradients might be fundamental for life anywhere.

Currently, the source of the organic carbon that led to the origin of life—the atmosphere, space, or hydrothermal vents—is unresolved. Given such a source, however, organic compounds must have been concentrated for chemical reactions to happen, which could have occurred when films formed on mineral surfaces, perhaps arising from drying or coldness. Then the simple organics could join up to make more complex molecules. But chemical systems need several other features before they can be considered 'alive'. The main ones are a metabolism, enclosure by a semi-permeable membrane, and reproduction that incorporates heredity with a genome.

With regard to the last characteristic, a more tractable issue for laboratory experimentation than the origin of life itself is a hypothesized stage of very early life, called the *RNA World*, in which RNA preceded DNA as the genetic material. In modern cells, DNA (deoxyribonucleic acid) provides an instruction set that is transcribed into RNA (ribonucleic acid) molecules that carry the information needed to make specific proteins. DNA is similar to RNA but more complicated, so it is logical that RNA evolved first. Moreover, in the 1980s it was discovered that some RNA molecules, known as ribozymes, can act as catalysts. So it is plausible that RNA once catalyzed its self-replication and assembly from smaller molecules. At the next stage, RNA would begin to make proteins, some of which would be better catalysts than RNA itself. Eventually, DNA would replace RNA because it is more stable and can be larger, both of which confer reproductive advantages.

Another critical feature in the origin of life was the encapsulation of a genome in a cell membrane. This was beneficial for a couple of reasons: first, to increase rates of reaction by concentrating biochemicals; and second, to give an evolutionary advantage to self-replicating molecules. For example, an RNA genome that produced a useful protein could keep it to itself inside a cell, thereby favouring its progeny.

So-called 'pre-cell' structures have been made in the lab. Spherical structures form spontaneously when oils are mixed with water, which is something we're all familiar with in the kitchen. Such oily or fatty membranes plausibly led to the membranes that surround modern cells and were surfaces for prebiotic chemistry.

Subsequent events may have occurred as follows. RNA took up residence in pre-cells, cells with an RNA genome evolved, and then modern cells with a DNA genome took over from RNA. At some point, metabolism also evolved, perhaps after RNA World. However, there remains an ongoing debate about which came first, metabolism or genetic replication. Also, the actual location, conditions, and evolutionary steps for the origin of life remain uncertain and are an area of research ripe for significant progress.

Chirality (or the art of clapping with one hand)

A description of life's origin would be incomplete without mentioning a related aspect of biochemistry: for biomolecules that come with two mirror-image structures, life on Earth uses only one of the structures, not both. The property of having mirror-image symmetry is called *chirality*, which comes from the Greek *cheir* for 'hand', given the fact that a person's hands are mirror images and cannot fit on top of one another (the same way up). If you're not convinced that a left-hand structure is truly different from a right one, try wearing your shoes on the wrong feet all day!

Chirality arises in biomolecules if a central carbon atom is surrounded by four different groups of atoms. The groups naturally arrange themselves in a tetrahedron, which is a pyramid with four triangular faces. If we number the groups with a '1' at the top of the tetrahedron, the three tetrahedral feet can be numbered clockwise or counterclockwise as shown:

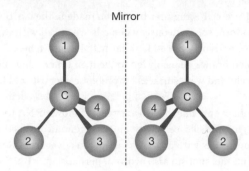

Mirror

The two molecules in the picture are 'handed' because one cannot be superimposed on top of the other. Such molecules are called *enantiomers* from the Greek *enantios morphe*, for opposite shape. Let's consider a specific enantiomer, the amino acid alanine. It would have '1' as a hydrogen atom, '2' as a carboxyl group (COOH), '3' as an amine group (NH_2), and '4' as a methyl group (CH_3). By convention, with the clockwise numbering of groups 2 to 4, alanine is said to be in the right-handed or D-form after the Greek *dextro* for right. With the counterclockwise arrangement, alanine is in the L form from the Greek *laevo* for left. Remarkably, life on Earth uses L-amino acids in proteins, almost entirely, and D-sugars almost exclusively as well whether in DNA, cell walls, or synthesis. (A handy mnemonic is 'lads' for left amino, dextro sugars). However, when a chiral substance is made in the laboratory, equal proportions of left- and right-handed forms are usually produced, called a *racemic* mixture.

Chiral molecules can have very different effects despite their identical chemical formulae. An example is the infamous drug thalidomide. The right-handed form cures morning sickness in pregnant women, but the left-handed version induces severe birth defects. A more pleasant example is the flavouring limonene: the D-form tastes like lemons but the L-form is orange-flavoured.

What caused the common chirality, or *homochirality*, shared by all life? There are two general hypotheses. One is that the organic molecules from which life started had a slight excess of one enantiomer that was subsequently amplified. Physical processes invoked to induce this excess include polarized light from stars that might cause organic molecules that seeded the Earth to develop a chiral bias during their synthesis in space; alternatively, polarized radiation from radioactive decay might generate an enantiomeric excess. A problem with such radiation-related ideas is that the excess is often small, less than 1 per cent. The second hypothesis is that prebiotic chemistry preferentially assembled only one of the enantiomers. For example, adsorption on a mineral surface might have chiral selectivity. Chemical kinetics is generally more efficient for a pure substance than a mixture, so this might cause further amplification. One biological process—reproduction of a genome—may require homochirality. Homochirality was probably needed for the replication in the RNA World model because only replicated RNA strands of the same chirality would line up with an RNA template strand.

Signs of the earliest life

Although the origin of life remains obscure, in the oldest sedimentary rocks on Earth we find possible signs of life. Perhaps it's no coincidence, but these rocks (mixed with volcanic flows) date from the end of the Late Heavy Bombardment, around 3.8 Ga. They are found in Isua, in the interior of south-west Greenland. They form a dark, foreboding landscape that looks something like Mordor, albeit icy, and they are remarkable for telling us much about the early Earth. There is gravel that was rounded by the action of water. Also, some sediments that were laid down under the sea contain flakes of carbon in the form of graphite. Isotopic analysis suggests that this carbon was once part of microbes living in the oceans. Carbon has two stable isotopes, carbon-12 and carbon-13. Life tends to concentrate a few per cent more carbon-12 atoms in its tissues than carbon-13 because molecules containing lighter carbon-12 atoms react faster in biochemistry. The graphite in Isua is enriched in carbon-12 by about 2 per cent, which is similar to the proportion in marine microbes. So the inference is that marine microbes died and sank into the sediments where the carbon was subsequently compressed and converted into graphite.

Better preserved evidence for early life is found in north-west Australia and dates from about 3.5 Ga. In the 1980s, my colleague Roger Buick found fossil stromatolites of this age near North Pole in north-western Australia (Fig. 2). An Aussie joker named the place 'North Pole' because it is one of the hottest, most sun-baked locations on Earth. Stromatolites are laminated sedimentary structures made by photosynthesizing microbial communities in water that is shallow enough to receive sunlight. Often stromatolites take the form of domes of wrinkly laminations. These domes are built up from trapped sediment as sheets of microbes, called *microbial mats*, growing upwards towards sunlight. On modern stromatolites, the living part—the mat—is just a veneer about a centimetre thick with a texture like tofu that sits atop accumulated

2. *Left*: Cross-section of the world's oldest fossil stromatolites in the Dresser Formation, North Pole, Australia. *Right*: A plan view of the bedding plane of the stromatolites, showing their tops. The lens cap is 5 cm diameter for scale

mineral layers. If you could travel back to the Archaean Earth, you would see coastlines covered in stromatolites all over the world. Today, stromatolites are rare because certain fish and snails (which didn't exist in the Archaean) feed on the microbial mats that make them. Consequently, modern stromatolites are found only in lagoons or lakes where the water is too salty for such voracious animals.

Some sceptics wonder if the North Pole structures could have formed without life and so are not really stromatolites, but most astrobiologists accept their organic heritage because they have various features we expect of biology. The fossil stromatolites have laminations with extremely irregular wrinkles that are difficult to explain as folds created by purely physical processes. Also, the laminations thicken at the top of convex flexures, which is just what happens when photosynthetic microbes grow where there is more sunlight. In troughs between flexures there are fragments that are plausibly interpreted as pieces of microbial mat that were ripped up by waves in the sea. The fragments contain thin, wrinkly layers that incorporate organic carbon. The combination of such observations suggests that the North Pole stromatolites are indeed the oldest fossil structures visible to the naked eye.

To find evidence for past life, we can also look for individual dead bodies of single-celled organisms, since only microbes existed so far back in Earth history. Such *microfossils*, which can only be seen with a microscope, are found in sedimentary rocks such as chert (a fine-grained silica-rich rock; flint is a familiar example) and shales, which are fine-grained sedimentary rocks that were once muds.

Geologists who study microfossils take rock samples from likely settings back to the lab where they slice them up and hope that something shows up under the microscope. It's a hit-and-miss affair. The problem of distinguishing life from non-life also rears its ugly head again because mineral grains sometimes look like fossil cells. Consequently, microfossils are more convincing if they show signs of cell division, colonial behaviour, or filamentous structures that occur when cells join up in a line. Microfossils can also be tested to see if they contain organic carbon.

The oldest incontestable microfossils occur in South Africa from 2.55 Ga. They are convincing because they are found in fossilized mats of filaments that contain organic carbon and include some microbial forms immobilized in the act of cell division. There are older possible microfossils in South Africa dated at 3.2–3.5 Ga that are spheroidal, contain organic carbon, and appear to show cell division, but have no other biological attributes. Similar microfossil colonies in 3.4 Ga sandstone of the Strelley Pool Formation in north-west Australia also possess organic carbon with isotopes characteristic of life.

The world's most controversial microfossils come from the Apex Chert rock formation in north-west Australia, with an age of 3.5 Ga. These were identified as the 'world's oldest microfossils' in the 1990s by Bill Schopf, of the University of California, Los Angeles. The structures appear as black to dark brown streaks that appear to be partitioned into cell-like sections. Schopf gave them Latin names implying that they were extinct cyanobacteria, which are photosynthetic bacteria. In 2002, a rancorous dispute arose when Martin Brasier of Oxford University re-examined the samples and showed that they didn't look like microbial filaments in three dimensions. Brasier also noted that the chert was not sedimentary but a type produced in a hydrothermal vent, which is an unlikely place to find cyanobacteria that need sunlight. Since then, further investigation has suggested that the filaments are really fractures that are partially filled with haematite, an iron oxide, which provides the dark colour. However, organic carbon is dispersed in the chert outside the filaments and the controversy continues.

So far, we've discussed isotopes, microfossils, and stromatolites. Further possible indicators of early life are *biomarkers*, which are recognizable derivatives of biological molecules. We're all familiar with the concept that fossil skeletons allow us to distinguish a *Tyrannosaurus rex* or *Triceratops*. On the microscopic scale, microbes can leave behind remnants of individual organic molecules, consisting of skeletal rings or chains of carbon atoms. These 'molecular backbones' come from particular molecules that are found only in certain types of microbe. So biomarkers not only confirm the presence of past life but can also indicate specific forms of life.

Unfortunately, like some microfossils, ancient biomarkers are also mired in controversy. The oldest reported biomarkers date from about 2.8 Ga and appear to show molecules suggesting the presence of oxygen-producing cyanobacteria. But the sampled rocks may have been contaminated by younger organic material. More certain biomarkers are found in rocks from 2.5 Ga inside tiny fluid inclusions that appear uncontaminated.

Despite the limitations of some evidence, the overall record shows that Earth was inhabited by 3.5 Ga—within 1 billion years of its formation and shortly after heavy impact bombardment, which suggests that life might originate fairly quickly on geological timescales on suitable planets. This means that it's not out of the question that Mars, which we think had a geologically short window of habitability (Chapter 6), may also have evolved life.

Chapter 4
From slime to the sublime

How did Earth maintain an environment fit for life?

However life started, once established, it persisted for over 3.5 billion years and evolved from microbial slime to the sophistication of human civilization. During this period, the Earth maintained oceans and, for the most part, a moderate climate, even though there was an increase of about 25 per cent in the amount of sunlight. The gradual brightening of the Sun is a consequence of the way that the Sun shines on the main sequence. When four hydrogen nuclei are fused into one helium nucleus in the Sun's core, there are fewer particles, so material above the core presses inwards to fill available space. The compressed core warms up, causing fusion reactions to proceed faster, so that the Sun brightens about 7–9 per cent every billion years. This theory is confirmed by observations of Sun-like stars of different ages.

If the Earth had possessed today's atmosphere at about 2 Ga or earlier, the whole planet should have been frozen under the fainter Sun, but geological evidence suggests otherwise. This puzzle is the *faint young Sun paradox*. There is evidence for liquid water back to at least 3.8 Ga, which includes, for example, the presence of sedimentary rocks that were formed when water washed material from the continents into the oceans.

There are three ways to resolve the faint young Sun paradox. The most likely explanation involves a greater greenhouse effect in the past. Another suggestion is that the ancient Earth as a whole was darker than today and absorbed more sunlight, although there's scant supporting evidence. A third idea is that the young Sun shed lots of material in a rapid outflow (solar wind) to space so that the Sun's core wasn't as compressed and heated over time by overlying weight as assumed above. If the mass loss was just right, the Sun could have started out as bright as today. However, observations of young Sun-like stars presently don't support the third hypothesis.

In considering the first idea, we need to appreciate that the atmosphere warms the Earth to an extent depending on atmospheric composition. Without an atmosphere (and assuming that the amount of sunlight that the Earth reflects stayed the same), the Earth's surface would be a chilly −18°C. Instead, today's average global surface temperature is +15°C. The 33°C difference (= 18 + 15) is the size of Earth's *greenhouse effect*, which is the warming caused by the atmosphere.

How does the greenhouse effect work? A planet's surface is heated by visible sunlight, causing it to glow in the infrared just as your warm body shines at those wavelengths. An atmosphere tends to be mostly opaque to the infrared radiation coming up from the surface below, so it absorbs the infrared energy and warms. Because the atmosphere is warm it also radiates in the infrared. Some of this radiation from the atmosphere travels back down to the planet. So the surface of a planet is warmer than it would be in the absence of an atmosphere because it receives energy from a heated atmosphere in addition to visible sunlight.

At 3.5 Ga, when the Sun was 25 per cent fainter, a 50°C greenhouse effect would have been needed to maintain the same global average temperature on Earth that we enjoy today with a 33°C greenhouse effect. A stronger greenhouse effect is possible if there had been greater levels of greenhouse gases, which are those responsible for absorbing infrared radiation coming from the Earth's surface. Today, water vapour accounts for about two-thirds of the 33°C greenhouse effect, and carbon dioxide (CO_2) accounts for most of the rest. However, water vapour condenses as rain or snow, so its concentration is basically a response to the background temperature set by atmospheric CO_2, which doesn't condense. In this way, CO_2 controls today's greenhouse effect even though its level is small. Around 1700, there were about 280 parts per million of CO_2 in Earth's atmosphere (meaning that in a million molecules of air, 280 were CO_2), while in 2010 there were 390 parts per million. Since industrialization, CO_2 has been released from deforestation and burning fossil fuels such as oil or coal. In the late Archaean, an upper limit on the CO_2 amount of tens to a hundred times higher than the pre-industrial level is deduced from the chemical analyses of *palaeosols*, which are fossilized soils. But even such levels of CO_2 were not enough to counter the faint young Sun.

In fact, the Archaean atmosphere had less than one part per million of molecular oxygen (O_2), which implies that methane (CH_4) was an important greenhouse gas. Today, atmospheric methane is at a low level of 1.8 parts per million because it reacts with oxygen, which is the second most abundant gas in the air at 21 per cent. (Most of today's air is nitrogen (N_2), which is 78 per cent, but neither oxygen nor nitrogen are greenhouse gases.) In the Archaean, the lack of oxygen would have allowed atmospheric methane to reach a level of thousands of parts per million. Methane wouldn't accumulate without limit because it can be decomposed by ultraviolet light in the upper atmosphere. Subsequent chemistry involving methane's decomposition fragments can generate other hydrocarbons, i.e. chemicals made of hydrogen and carbon, including ethane gas (C_2H_6) and a smog of sooty particles. A combination of methane, ethane, and carbon dioxide—plus the water vapour that builds up in response to the temperature set by these non-condensable greenhouse gases—would have provided enough greenhouse effect to offset the fainter Sun. This assumes that there was a source of methane from microbial life, just like today, which is plausible because the metabolism for methane generation is ancient (see Chapter 5).

Of course, we could ask what the Earth's early climate was like before life. In that case, CO_2 probably controlled the greenhouse effect, as today. In fact, throughout much of Earth history, there has been a geological cycle of CO_2 that regulated climate on timescales of about a million years. (It still operates today but is far too slow to counteract human-induced global warming.) Essentially, atmospheric CO_2 dissolves in rainwater and reacts with silicate rocks on the continents. The dissolved carbon from this *chemical weathering* then travels down rivers to the oceans, where it ends up in rocks on the seafloor, such as in limestone, which is calcium carbonate ($CaCO_3$). If deposition of carbonates were all that were happening, the Earth would lose atmospheric CO_2 and freeze, but there is a mechanism that returns CO_2 to the atmosphere. Seafloor carbonates are transported on slowly moving oceanic plates that descend beneath other plates in the process of *subduction*. An example today is the South Pacific 'Nazca' plate, which is sliding eastwards under Chile. Carbonates are squeezed and heated during subduction, causing them to decompose into CO_2. Volcanism (where rocks melt) and metamorphism (where rocks are heated and pressurized but don't melt) release CO_2. The whole cycle of CO_2 loss and replenishment is called the *carbonate–silicate cycle* and behaves like a thermostat. If the climate gets warm, more rainfall and faster weathering consume CO_2 and cool the Earth. If the Earth becomes cold, CO_2 removal from the dry air is slow, so CO_2 accumulates from geological emissions, increasing the greenhouse effect.

The carbonate–silicate cycle probably regulated climate before life originated. It then likely played an increasingly important role as a thermostat after levels of methane greenhouse gas declined in two steps when atmospheric oxygen concentrations increased, first around 2.4 Ga and then 750–580 Ma (Ma = millions of years ago).

A caveat is that the cycle may have operated differently in the Hadean and possibly the Archaean because *plate tectonics*—the large-scale motion of geologic plates that ride on convection cells within the mantle below—probably had a different style. Radioactive elements in the mantle generate heat when they decay, and more were decaying on the early Earth. So, on the one hand, a hotter, less rigid Hadean mantle should have allowed oceanic crust to sink more quickly. On the other hand, a hotter mantle should have produced more melting and thicker, warmer oceanic crust that was less prone to subduct. Overall, the presence of granites inferred from zircons (Chapter 3) implies that crust must have been buried somehow because granite is produced when sunken crust melts. However, exactly how tectonics operated on the early Earth remains an open question.

The Great Oxidation Event: a step towards complex life

The most drastic changes of the Earth's atmospheric composition have been increases in oxygen, which were just as important for the evolution of life as variations in greenhouse gases. For most of Earth's history, oxygen levels were so low that oxygen-breathing animals were impossible. The *Great Oxidation Event* is when the atmosphere first became oxygenated, 2.4–2.3 Ga. However, oxygen levels only reached somewhere between 0.2 to 2 per cent by volume, not today's 21 per cent. Large animals were precluded until around 580 Ma after oxygen had increased a second time to levels exceeding 3 per cent (Fig. 3).

Nearly all atmospheric oxygen (O_2) is biological. A tiny amount is produced without life when ultraviolet sunlight breaks up water vapour molecules (H_2O) in the upper atmosphere, causing them to release hydrogen. Net oxygen is left behind when the hydrogen escapes into space, thereby preventing water from being reconstituted. But abiotic oxygen production is small because the upper atmosphere is dry. Instead, the major source

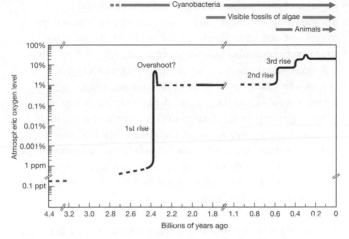

3. **The approximate history of atmospheric oxygen, based on geologic evidence (ppm = parts per million; ppt = parts per trillion)**

of oxygen is *oxygenic photosynthesis*, in which green plants, algae, and cyanobacteria use sunlight to split water into hydrogen and oxygen. These organisms combine the hydrogen with carbon dioxide to make organic matter, and they release O_2 as waste.

Another, more primitive, type of microbial photosynthesis that doesn't split water or release oxygen is *anoxygenic photosynthesis*. In this case, biomass is made using sunlight and hydrogen, hydrogen sulphide, or dissolved iron in hydrothermal areas around volcanoes. Today, microbial scum grows this way in hot springs.

Before plants and algae evolved, the earliest oxygen-producing organisms were similar to modern cyanobacteria. Cyanobacteria are bluish-green bacteria that teem in today's oceans and lakes. DNA studies show that a cyanobacteria-like microbe was the ancestor of plants, algae, and modern cyanobacteria. Consequently, we might suppose that the atmosphere became oxygenated once cyanobacteria evolved. However, evidence suggests that cyanobacteria were producing oxygen long before it flooded the atmosphere. A plausible explanation is that *reductants*, which are chemicals that consume oxygen, at first rapidly overwhelmed the oxygen. Reductants include gases such as hydrogen, carbon monoxide, and methane that come from volcanoes, geothermal areas, and seafloor vents. Distinctive iron-rich sedimentary rocks called *banded iron formations* dating from the Archaean show that there was considerable dissolved iron in the Archaean ocean, unlike today's ocean, which would have also reacted with oxygen.

The first signs of photosythetic oxygen appear about 2.7–2.8 Ga according to evidence from stromatolites and the presence of chemicals that became soluble by reacting with oxygen. In north-west Australia, stromatolites in rocks called the Tumbiana Formation once ringed ancient lakes. In theory, microbes might have built such stromatolites using anoxygenic photosynthesis, but there's no evidence for hydrothermal emissions needed for this metabolism. Instead, cyanobacteria using oxygenic photosynthesis probably constructed the stromatolites, consistent with tufts and pinnacles in the stromatolites that are produced when filaments of cyanobacteria glide towards sunlight. Furthermore, molybdenum and sulphur, which are elements only soluble when oxidized, become concentrated in seafloor sediments after 2.8 Ga to levels that are possible if microbes oxidized these elements on land using local sources of oxygen such as stromatolites.

Another reason why oxygen didn't immediately accumulate in the Earth's atmosphere is that its production is mostly a zero-sum process. When a molecule of oxygen is made, an accompanying molecule of organic carbon is generated, as summarized in the following equation:

carborn dioxide + water + sunlight = organic matter + oxygen

$$CO_2 + H_2O + \text{sunlight} = CH_2O + O_2$$

The process easily reverses, i.e. organic matter reacts with oxygen to regenerate carbon dioxide and water. Also, the Archaean atmosphere's lack of oxygen meant that microbes could readily convert organic carbon into methane gas, as happens today in smelly, oxygen-free lake or seafloor sediments. In the air, methane could react with the oxygen to recreate water vapour and carbon dioxide. In both cases—direct reversal or indirect cancellation with methane—no net oxygen is produced, despite the presence of photosynthesis.

However, a tiny fraction (about 0.2 per cent, today) of organic carbon is buried in sediments and separated from oxygen, which prevents the two from recombining. Every organic carbon that gets buried is equivalent to one net O_2 molecule that's freed up. Of course, this 'freed' oxygen can easily react with many other substances besides organic carbon, including geological gases and dissolved minerals such as iron. Nonetheless, a tipping point was reached when the flow of reductants to the atmosphere dropped below the net flow of oxygen associated with organic carbon burial, causing the Great Oxidation Event. Afterwards, a plateau of 0.2–2 per cent oxygen was reached probably because oxygen that dissolved in rainwater reacted significantly with continental minerals, preventing its further accrual.

Indeed, a surge in oxidation is the evidence for the Great Oxidation. 'Red beds'—rust-coloured continental surfaces that arise when iron minerals react with oxygen—appear. Today, reddish land surfaces are common, such as in the American south-west. A change in sulphur processing in the atmosphere also occurred. Before the Great Oxidation, when there was less than one part per million of atmospheric oxygen, red- and yellow-coloured particles of elemental sulphur literally fell out of the sky. They formed in chemical reactions at many kilometres altitude when sulphur dioxide gas from volcanoes was broken up by ultraviolet light. The falling sulphur particles carried a sulphur isotope composition into rocks that indicates atmospheric formation. After the Great Oxidation, oxygen combined with atmospheric sulphur, preventing elemental particles from forming. The sky cleared and the isotopic signature vanishes from sedimentary rocks.

With the Great Oxidation, Earth's stratospheric ozone layer formed. This region of concentrated ozone between 20 and 30 kilometres in height shields the Earth's surface from harmful ultraviolet rays. If there were no ozone layer, you would be severely sunburned in tens of seconds. Ozone is a molecule of three oxygen atoms (O_3), which comes from oxygen. An oxygen molecule (O_2) is split by sunlight into oxygen atoms (O), which then combine with other O_2 molecules to make ozone (O_3).

The cause of the Great Oxidation is debated but must boil down to either a global increase in oxygen production from more organic burial or a decrease in oxygen consumption. The problem with the first idea is that organic matter extracts the light carbon isotope, carbon-12, from seawater, but seawater carbon, recorded in ancient limestones, doesn't steadily decrease in carbon-12 content. That leaves the second idea.

But what would have caused the flow of reductants to diminish? One possibility is that the relatively large abundances of hydrogen-bearing gases such as methane and hydrogen (which are inevitable in an atmosphere without oxygen) meant that hydrogen escaped rapidly from the upper atmosphere into space. The loss of hydrogen oxidized the solid planet. Oxidized rocks have fewer reductants so this gradually throttled further release of reductants. Oxidation happens because you can generally trace escaping hydrogen back to its source in water (H_2O), even if hydrogen atoms go through methane (CH_4) or hydrogen (H_2)

intermediaries. So when hydrogen leaves, oxygen gets left behind where the hydrogen originated. Water vapour itself has trouble reaching the upper atmosphere because it condenses into clouds, but other hydrogen-bearing gases such as methane don't condense and sneak the hydrogen out into space.

A rough analogy for the Great Oxidation is an acid-base titration of the sort performed in high school by dripping acid into an alkaline solution. Suddenly, the solution changes colour, typically from clear to red. The ancient atmosphere reached a similar transition from hydrogen rich to oxygen rich. Rather than acid base, we call such a titration a '*redox* titration' because it's a competition between *re*ductants such as hydrogen and *oxi*dants such as oxygen.

A boring billion years ended by the advent of animal life

The ocean and land adjusted to the change of the Great Oxidation, and by about 1.8 Ga, atmospheric oxygen had settled down and remained between 0.2 and 2 per cent for an amazingly long time. The evolution towards complex life was slow perhaps because anoxic waters that were often rich in dissolved iron or hydrogen sulphide underlay a moderately oxygenated surface ocean. Such anoxic conditions are toxic for complex life. In fact, evolution was so sluggish that the interval from 1.8 to 0.8 Ga is called the *boring billion*. According to one scientific paper, 'never in the course of Earth's history did so little happen to so much for so long'!

Actually, several notable events did occur in the otherwise boring billion and just before it. Around 1.9 Ga, the first fossils of single organisms appear that are visible to the naked eye. These impressions of spiral coils of a few centimetres diameter (named *Grypania*) were probably seaweeds. In China, *acritarchs*, a type of fossil larger than 0.05 mm in size with organic-walls, occur at 1.8 Ga. Some are

53

thought to have been algal cysts, which form when algae turn into inactive balls during dry periods. New lineages also emerged, including red algae at 1.2 Ga, which were likely capable of sexual reproduction. Today, we boil red algae descendants to make agar, which is used to thicken ice cream, and we use one type, nori, to wrap sushi. Because sex was discovered, the boring billion was not *that* boring.

Starting around 750 Ma, and culminating after fits and starts about 580 Ma, atmospheric oxygen rose a second time to levels exceeding 3 per cent, and the deep sea became fully oxygenated. Animal biomarkers and tiny fossils (possibly sponges) date from around 630 Ma. But it's only after 580 Ma that we find complex fossils from centimetres to metres in size. These are strange soft-bodied organisms without mouths or muscles that must have relied upon diffusion of nutrients through their skin. Some resemble pizzas while others look like plant-like fronds. However, the organisms lived too deeply in the sea to receive sunlight so they cannot have been plants. Collectively, they are called the *Ediacaran biota* because they were first clearly identified in the Ediacaran Hills in southern Australia. The Ediacarans lasted tens of millions of years before dying out.

Then, the *Cambrian Explosion* occurred, which was the relatively rapid appearance of animal fossils with new body architectures in a 10- to 30-million-year interval after the beginning of the Cambrian Period at 541 Ma. Representatives of most modern groups of animals emerged, including many with hard skeletons. The Cambrian (541–488 Ma) comes directly after the end of Proterozoic Aeon, and is the first period of the Phanerozoic Aeon (541 Ma to present). Phanerozoic means 'visible animals' from the Greek *phaneros* (visible) and *zoion* (animal). One tiny Cambrian creature, *Pikaia*, was a sort of swimming worm, a few centimetres long. *Pikaia* was probably in a group from which all the vertebrates, including us, evolved.

The second rise of oxygen allowed a diversity of animals to arise but its cause remains unsolved. Suggestions include organic carbon burial (and hence oxygen production) accelerated by several factors. These were new life on land that boosted the breakdown of surfaces and nutrient release, the evolution of marine zooplankton, and the enhanced production of clays that helped to bury organic matter. Alternatively, moderate levels of around 100 parts per million of atmospheric methane throughout the 'boring billion' might have promoted yet more hydrogen escape to space that oxidized the seafloor and ushered in an increase of oxygen after a second global redox titration.

Around 400 Ma, a third rise of oxygen took place when vascular plants colonized the continents. Such plants have special tissues (such as wood) to transport water and minerals. They probably enhanced nutrient release (by breaking down land surfaces with roots) and organic burial. Afterwards, atmospheric oxygen levels remained within 10 per cent to 30 per cent, allowing the persistence of animals and eventually our own appearance.

Snowball Earth or Waterbelt Earth?

Curiously, worldwide glaciations are associated with both the Great Oxidation Event and the second rise of oxygen, like bookends encompassing the 'boring billion'. Some scientists argue that at these times, the Earth's oceans entirely iced over in a so-called *Snowball Earth*. This state was far more extreme than the ice ages that have occurred in the last few million years that have only extended down to mid-latitudes, not the tropics. For context, the Proterozoic aeon is divided into three eras, the Palaeoproterozoic (2.5–1.6 Ga), Mesoproterozoic (1.6–1.0 Ga), and Neoproterozoic (1.0–0.541 Ga). Geologists commonly speak of the 'Palaeoproterozoic glaciations' to describe three or four glaciations that occurred at the beginning (2.45–2.22 Ga) of the Palaeoproterozoic, while the 'Neoproterozoic glaciations' refers to two large glaciations (at 715 and 635 Ma) and one small one (at 582 Ma).

The rock record provides evidence of these worldwide glaciations. The flow of ice sheets mixes stones, sand, and mud into a type of rock called *tillite*. Glaciers also leave parallel scratch marks called striations on underlying rocks. Furthermore, rocks drop out of icebergs or melting ice shelves and end up below as so-called *dropstones* that deform sedimentary beds. Together, tillites, *glacial striations*, and dropstones are dead giveaways that ice sheets were once present.

Magnetic measurements made in the 1980s and 1990s also demonstrated that the glaciations extended into the tropics. A compass shows the Earth's north–south magnetic field. Less obvious is the fact that the magnetic field lines are approximately perpendicular to the surface near the poles and parallel to the surface near the equator. Volcanic rocks contain iron minerals, such as magnetite (an iron oxide, Fe_3O_4), which become magnetized in the direction of the prevailing magnetic field when they cool. Thus, when the trapped or remnant magnetic field lines are parallel to ancient beds of rock, we know that the rocks formed in the tropics. This was discovered in rocks associated with Neoproterozoic and Paleoproterozoic glaciations.

The idea that the entire Earth might freeze over if sea ice reached the tropics originated in the 1960s. A Russian climatologist, Mikhail Budyko, calculated that if polar ice caps extended farther equatorward than 30° latitude, the entire Earth should freeze. Ice reflects a lot of sunlight and so has a high *albedo,* which is the fraction (between 0 and 1) reflected. Today, the Earth's albedo is about 0.3, meaning that 30 per cent of the sunlight is reflected to space. But sea ice reflects 50 per cent when bare or 70 per cent if covered in snow. A disastrous *ice-albedo runaway* occurs when high-albedo ice in the tropics makes the Earth absorb less sunlight and cool, which creates more ice that makes the Earth even colder, and so on. A frozen globe is created with an average temperature below –35°C and an ice thickness of 1.5 km at the equator and 3 km at the poles.

Joe Kirschvink of the California Institute of Technology proposed how a Snowball Earth would melt, as well as the very phrase *Snowball Earth*. He suggested that volcanoes would punch through the ice and vent carbon dioxide into the atmosphere. Because there would be no rainfall or significant open ocean to dissolve carbon dioxide, it would build up over a few million years until an enhanced greenhouse effect melted the Snowball. Also, the flow of ice from thick polar ice to thin tropical ice would carry volcanic ash and windblown dust, so that the tropics would become darker, making the Snowball prone to melt. Then the ice-albedo runaway would run in reverse: less ice would make the Earth warmer, which would melt more ice, and so on.

Layers of carbonate rock called *cap carbonates* that sit on top of the glacial deposits possibly attest to this aftermath of a Snowball Earth. The idea is that after the ice melted, large amounts of carbon dioxide were converted into cap carbonates. Proponents of the Snowball Earth hypothesis note that the ratio of carbon-12 to carbon-13 isotopes in the carbonate is similar to that in volcanic carbon dioxide.

A central problem for the Snowball Earth hypothesis is how photosynthetic algae survived under the thick ice, given that they require sunlight. Lineages of green and red algae went straight through the Neoproterozoic snowballs unscathed. One suggestion is that heat around volcanoes allowed some open water or thin, transparent ice. But others question whether the entire Earth really froze over. Some reasonably sophisticated computer simulations of the Earth's ancient climate produce a tropical belt of open water—a *Waterbelt Earth*—because this region is surrounded by bare ice of moderate albedo, whereas the very reflective, snow-covered ice occurs only at high latitudes.

Whatever happened, the cause of both the Neoproterozoic and Paleoproterozoic glaciations was probably a diminished greenhouse effect. Continents drift because of plate tectonics, and reconstructions have them bunched up in the tropics prior to the ancient snowball eras. Rainfall over tropical continents perhaps drew carbon dioxide down to lower levels than typical, poising the Earth for glaciation. Also, the coincidence of rises in oxygen with the glaciation eras suggests a trigger through methane, another greenhouse gas. Probably the oxygen increases were associated with rapid decreases in methane because high abundances of the two gases are incompatible—they react.

For astrobiology, the lesson from Snowball Earth is that the biosphere on an Earth-like planet is resilient. Despite drastic climate swings, life on Earth survived and emerged after the Neoproterozoic into a new aeon of animals.

On the occurrence of advanced life

In astrobiology, we often wonder if animal-like life exists elsewhere in the galaxy. That possibility can perhaps be informed by evolution on Earth. To become complex, terrestrial life had to overcome at least two key hurdles. First, life had to acquire complex cells, which were needed for differentiation on the larger scale, i.e. cells with functional distinctions, such as liver versus brain cells. Large, three-dimensional multicellular animals and plants are made only of *eukaryotic* cells, which can develop into a much larger range of cell types compared to the cells of microbes (see Chapter 5).

A second precursor for animals was having sufficient oxygen to generate lots of energy with an *aerobic* (oxygen-using) metabolism, such as our own. Abundant energy is needed to grow large and move. *Anaerobic* metabolism, which doesn't use oxygen, produces about ten times less energy for the same food intake than aerobic metabolism. For extraterrestrial life, we might wonder if fluorine or chlorine—which are powerful oxidants—could be used in place of oxygen to generate high energy levels. The answer is no. Fluorine is so reactive that it explodes with organic matter, while chlorine dissolves to make harmful bleach. Oxygen, it appears, is the best elixir for complex life in the periodic table—reactive enough but not too aggressive.

Oxygen not only needs to be present but also concentrated. The first aerobic multicellular life probably consisted of aggregates of cells produced when dividing cells failed to separate. Such agglomerations would have been limited in size by diffusion of oxygen to the innermost cell so that greater levels of external oxygen would have permitted larger cell groupings. At typical metabolic rates, a diffusion-constrained animal precursor of about a few millimetres' dimension would need atmospheric oxygen concentrations exceeding 3 per cent. This corresponds to the oxygen increase around 580 Ma that ushered in the animals.

Fossil algae show that eukaryotes existed long before animals, so the slow fuse to the Cambrian Explosion was probably the time it took to build up adequate oxygen. We can even define an *oxygenation time* as that required to reach oxygen levels sufficient for complex animal life. On Earth, it was 4 billion years after the planet formed.

The oxygenation time on Earth-like exoplanets is uncertain given that we're still trying to understand what set the timescale for the Earth. However, oxygen comes from liquid water. So Earth-like exoplanets with oceans have the potential to develop oxygen-rich atmospheres if water-splitting photosynthesis evolves. If the oxygenation time exceeds 12 billion years on a certain exoplanet because it's endowed with material that reacts with oxygen, a Sun-like parent star would turn into a 'red giant' before the conditions for animal life became possible, and the planet would be doomed to possess nothing more complex than microbial slime. But if other exoplanets have short oxygenation times, complex life might occur faster than it took on Earth.

Trends in evolution? The lesson of mass extinctions

Once animals evolve, we might wonder if evolution has a trend towards greater complexity. Consider life on land. In land colonization, the first plant spores are found around 470 Ma, while fossils of substantial parts of plants appear about 425 Ma. Insects appear on land around 400 Ma. Then fish evolved into amphibians around 365 Ma and their descendants ultimately became the reptiles and mammals, including us. This is a trend for the biosphere to use more of the Earth's resources. But a misconception is to think of evolution as a steady progression ending in us. The fossil record shows that species come and go. Also, mass extinctions intermittently prune the diversity of complex life. They are events when more than 25 per cent of families are lost, where 'family' is the biological tier above genus and species (see Chapter 5). Such mass extinctions refer to the non-microbial part of the biosphere, of course.

Only around the Proterozoic–Phanerozoic transition is there a dramatic increase in diversity because at that time organisms acquired new body architectures and ways of living that persisted. In animals, a key innovation was the body cavity, the *coelom* (pronounced 'see-lum'), which could be filled with fluid to provide rigidity and allow the concentration of forces from muscles. This facilitated self-propulsion, which was further improved with hard skeletons, at first external and then internal. A predator–prey evolutionary arms race likely contributed to the Cambrian Explosion of animal varieties. Plant diversity later increased with the development of vascular systems with rigid cells for transport and associated organs that eventually gave rise to trees.

In the past 500 Ma, there have been five mass extinctions that killed over 50 per cent of species, with the two biggest at 251 Ma and 65 Ma, respectively. Excluding crocodilians, no land animal larger than the size of a domestic dog got through the largest event, separating the Permian (299–251 Ma) and Triassic

(251–200 Ma) periods. This event took a massive toll in the oceans. For example, the trilobites—icons of the Cambrian seas—had been in decline but the Permian–Triassic extinction finished them off. The second largest event at 65 Ma wiped out the dinosaurs and separates the Cretaceous (145–65 Ma) and Paleogene (65–23 Ma) periods. (Some literature refers to this event as Cretaceous-Tertiary based on an older naming system.) The Permian–Triassic mass extinction destroyed as much as 95 per cent of marine species and 70–80 per cent of land animals, while the Cretaceous–Paleogene extinguished 65–75 per cent of all species.

The cause of the Permian–Triassic mass extinction was apparently a chain of misfortunes generated by the Earth herself. The trigger seems to have been large-scale volcanism in Siberia, covering an area similar to Europe. Because coal deposits underlay this area, the volcanism pumped out huge amounts of coal-derived methane and carbon dioxide greenhouse gases, which warmed and acidified the ocean. Oxygen is less soluble in warm water, so an anoxic deep sea developed, which may have belched poisonous hydrogen sulphide to the surface. The combination of climatic warming, ocean acidification, and noxious gases extinguished more life than any other event in the Phanerozoic.

The impact of an asteroid of about 10 kilometres diameter appears to have caused the Creataceous–Paleogene mass extinction. Calamitous consequences included temporary destruction of the ozone layer, worldwide wildfires, acid rain, and subsequent climatic cooling caused by sulphate particles injected into the stratosphere that reflected sunlight. The impact crater has even been found about a kilometre underground near the Mexican town of Chicxulub (pronounced 'sheek-soo-loob').

Extinctions destroy previously successful lineages but they also provide opportunities for others. You are reading this book because of the Chicxulub impactor. The mammals became dominant once the dinosaurs were gone.

On the other hand, mass extinction events mean that if a civilization develops, it can be wiped out. Impacts the size of that producing the Cretaceous–Paleogene extinction should occur every 100 million years or so on Earth, give or take a factor of a few. The implication is that civilizations on exoplanets could be short-lived compared with the age of the universe merely because of such random catastrophes. Thus, an Earth-like planet can remain fit for life, but civilizations are probably ephemeral. The latter has consequences in the astronomical search for extraterrestrial intelligence (Chapter 7).

Chapter 5
Life: a genome's way of making more and fitter genomes

Life on Earth: the view from above

In Chapter 1, I presented a general definition of life, but in finding life elsewhere, it helps to know our one example of life in great detail. To this end, let's start with a global perspective of terrestrial biology and work downward to cells and molecules.

Imagine an interstellar traveller who arrives on the Earth and wants to know about our biology. Perhaps from her planet, she had deduced that life exists here because of Earth's wet, anomalously oxygen-rich atmosphere, or she had picked up TV broadcasts from decades ago that somehow didn't put her off visiting. Amazingly, she speaks English (as extraterrestrials always do in the movies) and by a stroke of luck her spacecraft lands in an English-speaking country! What would we tell her?

At the global level, Earth's *biosphere* is the sum of all living and dead organisms. Sometimes the term includes the non-living regions that life occupies. By quantifying the biosphere in billions of tonnes of carbon (1 tonne = 1,000 kg), we can identify its broad components. The biomass on land is around 2,000 billion tonnes of carbon of which 30–50 per cent is living and the rest is dead. In the ocean, only 0.1–0.2 per cent of about 1,000 billion tonnes of biomass carbon is alive. Forests are the reason that there's so much more living biomass on land than in the oceans.

An unsettled issue that's relevant to life elsewhere (below the surface of Mars or Jupiter's moon, Europa) is the extent of Earth's *subsurface biosphere* or 'intraterrestrial life.' Some scientists suggest that a huge mass of microbes extends a kilometre or two below the seafloor and more than 3 kilometres underneath land. A limit for life at such depths is temperature. As you go downwards, it gets warmer (as miners know) and, at some point, too hot for even the toughest microbes. Earth's subsurface biosphere biomass is uncertain because deep drilling has not been done for all types of subsurface environments, but estimates range from about 1 per cent to 30 per cent of Earth's living biomass.

Whatever its precise mass, the biosphere is less than a billionth of the mass of the Earth and yet manages to greatly influence the chemistry of the surface environment. The biosphere can do this because it processes vast amounts of material with rapid turnover. In doing so, individuals live and die almost instantly on geological timescales. Microbes typically reproduce in tens of minutes to days, while large multicellular organisms last only a few thousand years at most before they become dead fodder for microbial degradation. For example, amongst the latter, the oldest non-clonal organism is a Bristlecone Pine in the White Mountains of California, which germinated around 3049 BC, several centuries before the earliest Egyptian pyramids, but a mere blip in geological time.

The main activities in today's biosphere are oxygenic photosynthesis and its chemical reversal through aerobic respiration and oxidation (Chapter 4). However, microbes possess a huge range of other metabolisms, which we discuss later.

Microbes, of course, are found almost everywhere at the Earth's surface. A seawater and freshwater bacterium called *Pelagibacter ubique* is probably the most numerous organism on Earth. Despite that, it was first described only in 2002, which shows how biology is still developing. Overall, there are about 10^{29} microbes of all varieties in the ocean, far exceeding the 10^{22} stars in the observable universe. Microbes are also abundant on land. There are typically about 100 million to 10 billion microbes per gramme of topsoil. Indoor air usually has about a million bacteria per cubic metre. An average of one microbe (albeit dead) even floats in every 55 cubic metres of air at 32 kilometres altitude in the stratosphere.

There are four key properties that have allowed the Earth to become so extensively inhabited. The most important is widespread liquid water. All metabolizing organisms contain organic molecules dispersed in aqueous solution. Consequently, we don't expect life to exist on a completely dry surface, such as Venus. The second essential property is having energy for metabolism. Sunlight is the main source for Earth's biosphere, but some organisms obtain energy from chemical reactions in darkness, which means that it's not impossible that microbial-like life might exist below the surface of Mars or Europa. A third life-giving feature is a renewable supply of essential chemical elements. A planet that can't resupply vital elements through natural cycles (such as the water cycle or tectonics) would be deathly. A fourth attribute, which may be essential for life, is the presence of interfaces between solids, liquids, and gases. It's advantageous to live at a stable interface, such as on land or the surface of the ocean. This is why it would be difficult for life to exist on a gas giant like Jupiter that has no surface. Life might speculatively be possible at certain altitudes in Jupiter's atmosphere but deep churning by convection would periodically plunge life into an interior of fatal heat and pressure.

An inside view of terrestrial life: the cell

We have just considered biology on a global scale but under the microscope all organisms are composed of cells of different types that put them into one of three domains, the *Eukarya*, *Archaea*, or *Bacteria*. The last two are microbial and are sometimes lumped together as *prokaryotes*, although many microbiologists now consider this term antiquated because archaea and bacteria are biochemically dissimilar. DNA floats freely in the middle of the archaeal and bacterial cells, whereas in eukaryotes the DNA is housed inside a membrane-bound nucleus. Archaea and bacteria are single cells, with the exception of some bacterial species that join up in a row to form filaments. Eukaryotes can be single celled, such as an amoeba or a baker's yeast, but only eukaryotes form large, three-dimensional multicellular organisms, such as mushrooms or humans.

The classification into three domains of life was motivated by genetics and supersedes an older 'five kingdom' system of plants, animals, fungi, protists (single-celled eukaryotes), and bacteria. However, these old terms are still used in taxonomy, which classifies an organism below its domain according to Kingdom, Phylum, Class, Order, Family, Genus, and Species. The mnemonic I use to remember these levels (or taxa) is far too rude to mention but another is 'Keeping Precious Creatures Organized For Grumpy Scientists'. The taxonomic levels of a human, for example, are the animal kingdom, chordate phylum, mammal class, primate order, hominid family, *Homo* genus, and *sapiens* species. The Swedish botanist Carolus Linnaeus (1707–78), who gave us the word *biology*, developed modern taxonomy, including binomial names for organisms, e.g. *Homo sapiens*. But it took the advent of evolutionary theory and molecular biology to uncover the biochemical unity of life and genetic common ancestry.

While there are some similarities between eukaryotes and the other two domains, there's also a gulf in complexity. Bacteria and archaea are usually around 0.2–5 microns (millionths of a metre) in size, with rare exceptions, whereas eukaryotic cells are generally bigger, at 10–100 microns size. The larger eukaryotic cells contain organelles to perform specialized functions, analogous to the

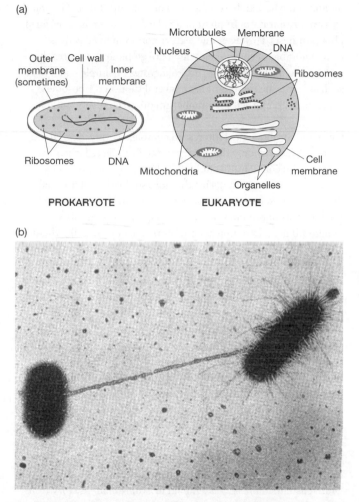

4. a) Schematic of prokaryote (archaea and bacteria) versus eukaryote structure; b) Two bacteria caught in the act of conjugation

organs of the human body (Fig. 4). For example, the *mitochondria* carry out respiration. In plant or algal cells, *chloroplasts* perform photosynthesis. However, one feature common to all cells is a large number of ribosomes, which are globular structures that make proteins. For example, in prokaryotes or simple eukaryotes such as yeast, the cell might have several thousand ribosomes, while in an animal cell the number might reach several million.

Given an astrobiological interest in complex extraterrestrial life, we might ask why only the eukaryotic cell produces large, three-dimensional multicellular life. The answer is not fully known but eukaryotic cells have a more sophisticated internal dynamic cell skeleton or cytoskeleton than archaea or bacteria. This consists of protein microfilaments, tiny protein tubes ('microtubules'), and molecular motors that control cell structure and help transport signalling molecules to change cell physiology. So the ability to develop into many specialized forms (for example, skin or brain cells) is inherent in the make-up of a eukaryotic cell. Cyanobacteria are able to make filaments of hundreds of cells in a row with some different cell types but that is the limit. Unlike archaea or bacteria, eukaryotes have bigger and more modular genomes, which also allows for more complexity. If there were no eukaryotic cells, the Earth would be much duller. All of the familiar organisms of our world—the animals, plants, and fungi—wouldn't exist. Thus, when we think about complex life on exoplanets, we should wonder whether evolution would make cells like eukaryotes elsewhere. For this reason, the origin of eukaryotes is of great interest.

Modern genetics implies that the eukaryotic cell is a 'Frankenstein's monster', assembled in evolutionary history from bits and pieces of bacteria and archaea. For example, the mitochondrion in eukaryotic cells was derived from a bacterium originally living symbiotically inside another cell. The larger cell, which may have been an archaeon or some 'proto-eukaryote' that no longer exists, swallowed the free-living bacterium and, in the most important gulp in history, the mitochondrial ancestor came into being. Indeed, separate DNA in mitochondria provides evidence of bacterial ancestry. The distinct DNA of chloroplasts in plant and algal cells shows that chloroplasts were derived in a similar way from symbiotic cyanobacteria that ended up living inside larger cells. Effectively, all the cells in the leaves of green plants contain ancestors of cyanobacterial slaves that were caged and co-opted long ago. The theory of such an origin for the mitochondria, chloroplasts, and other organelles in eukaryotic cells is called *endosymbiosis*.

A world without eukaryotes would also be one without sex. Think no flowers or love songs. Archaea and bacteria are not sexed but they do conjugate when two cells are connected by a tube in which genes are transferred in pieces of DNA called plasmids. Bacterial surfaces have protuberances called pili, and during conjugation a special pilus extends to a partner, providing the conduit (Fig. 4). Unlike sexual reproduction in eukaryotes, microbial conjugation doesn't produce offspring and is quick and easy. It's as if you brushed up against a person with red hair in a coffee shop and acquired the red-haired gene with an instant change of your hair colour. Archaea and bacteria can also acquire new genes through transformation (uptake of foreign DNA from the environment) and transduction (virus-mediated gene swapping). Indeed, the rapid acquisition of genes allows the quick development of bacterial resistance to antibiotics.

Eukaryotes that are sexed generate *gametes*, i.e. sperm and egg cells. In complex multicellular eukaryotes, the diversity of life makes it surprisingly hard to define 'male' and 'female', leaving us with the odd definition that males are those that produce the small gametes, while females produce the big ones. The gametes fuse so that half of the genes come from a father and half from a mother. Usually DNA is a loosely coiled thread but before cell division, DNA curls up into visible *chromosomes* under a microscope. For example, humans have 46 chromosomes in 23 pairs in each cell, except for the *gamete*s, which have half, i.e. 23 chromosomes.

There are many ideas about why sex is evolutionarily advantageous for eukaryotes. One possibility concerns how it mixes and matches genes from both parents onto each chromosome in a process called *recombination*. If beneficial mutations occur separately in two individuals, the mixture of both can't be achieved in asexual organisms, but sexually reproducing organisms can bring them together and reap the benefits. Conversely, sex can also eliminate bad, mutated genes by bringing unmutated genes together in some individuals, whereas self-cloning organisms are stuck with bad genes, and offspring can die because of them.

Outside the three domains, viruses represent a grey area between the living and non-living. Viruses are typically about ten times more abundant than microbes in seawater or soil. They consist of pieces of DNA or RNA surrounded by protein and, in some cases, a further membrane. Viruses are tiny, only about 50–450 nanometres (billionths of a metre) in size, comparable to the wavelength of ultraviolet light. They are generally considered non-living because they are inanimate outside a cell and have to infect and hijack cells for their own reproduction. However, some do this without the host ever noticing, so not all viruses cause disease. One theory of several for the origin of the nucleus of eukaryotes is that it may have evolved from a large DNA virus, but the role of viruses in the evolution of life is still a matter of debate.

The chemistry of life

To discuss many aspects of life, such as genetics and metabolism, requires the vocabulary of biochemistry. The four main classes of biomolecule are nucleic acids, carbohydrates, proteins, and lipids.

Like self-assembly furniture, many biomolecules are modular.
They are chains or polymers of smaller units called monomers.

Carbohydrates provide energy and structure. They contain C, H,
and O atoms in a 1:2:1 ratio, with repeated units of $C(H_2O)$, so
literally, their chemical composition is 'carbon hydrated'. Sugars
with five carbon atoms are found in DNA and RNA molecules,
while six-carbon sugars exist in cell walls, such as cellulose in
plants.

Lipids are organic molecules that are insoluble in water but
dissolve in a non-polar organic solvent—one without significant
electrical charge on any of its atoms, such as olive oil. Thus, if we
ground up a dead animal and bathed it in a non-polar solvent,
anything that dissolved would be a lipid. Life uses lipids in cell
membranes, as fats for energy storage, and as signalling
molecules. Major components of membranes are phospholipids,
which have a hydrophilic (water-loving) end that contains
phosphorus and a hydrophobic (water-repellent) tail that is a
hydrocarbon, consisting of carbon and hydrogen atoms. A double
layer of phospholipids, called a *bilayer*, forms a membrane. The
hydrophilic ends stick out into an aqueous medium on the interior
and exterior of a cell, while hydrophobic tails face each other in
the middle of the membrane.

Proteins are polymers made up of amino acid units. Their use
includes enzymes and structural molecules, although there is a
long list of other functions.

RNA and DNA are nucleic acids, which are polymers of *nucleotide
monomers*. Each nucleotide is made up of a five-carbon sugar, a
phosphate, and a part called a base (Fig. 5). In DNA, there
are four possible bases. All of them contain one or two rings of six
atoms where four of the atoms are carbon and two are nitrogen.
Each base has a letter designation of A, C, G, and T, which
stand for adenine, cytosine, guanine, and thymine molecules,

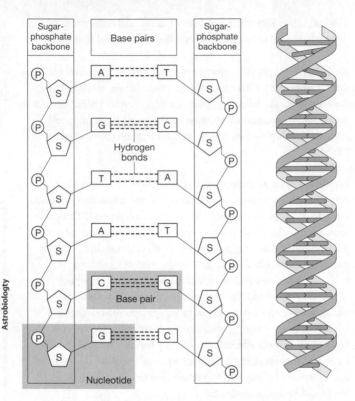

5. *Left*: DNA consists of two strands connected together. Each strand is made up of a 'backbone' of phosphate (P) and sugar (S) components. The strands connect to each other with base pairs. *Right*: In three dimensions, each strand is a helix, so that overall we have a 'double helix'

respectively. Molecules of RNA use the same three bases except that the T of DNA is replaced by a U for uracil.

In 1953, James Watson and Francis Crick famously deduced the structure of DNA: two polynucleotide strands coiled in a screw-like helix (Fig. 5). The bases at the sides of each strand stick together by 'hydrogen bonds' in which a slightly positively charged hydrogen atom on one base is attracted to a slightly negatively charged atom on a base from the opposite strand. Structural compatibility only allows a C base to pair with G and an A to pair with T. Each DNA molecule has several million nucleotides, and each chromosome in the cell contains a DNA double helix. In contrast, RNA molecules are mostly single stranded. However, RNA can fold back on itself if complementary bases exist in two separate parts of the strand, noting that adenine pairs with uracil (U) in RNA.

The structure of DNA incorporates two fundamental characteristics of life identified in Chapter 1: an ability to reproduce; and a blueprint for development and maintenance. In replication, the DNA helix splits into two strands with each serving as a template for a new complementary strand. For example, wherever A appears on the template, a T is added to the newly generated strand, or vice versa. The same applies for G–C pairs. In the process, mutations and mistakes allow for evolution.

In fulfilling its other role as the blueprint, the DNA double helix is unzipped by an enzyme to provide instructions to generate proteins. Part of the unzipped DNA undergoes *transcription* into a strand of messenger RNA (mRNA), which is a complementary copy of the DNA, except that U (instead of T) is inserted wherever A appears in the DNA. Then the mRNA is fed into a ribosome like a ribbon. In the *genetic code*, groups of three letters (called *codons*) along the mRNA specify each amino acid in a protein that flows out of the ribosome.

Apart from reproduction, life has to sustain itself through metabolism, which involves breaking down molecules to make energy (*catabolism*) as well as building up biomolecules (*anabolism*). The classification of metabolisms depends on the need for energy and carbon. Because each of these requirements can, in turn, be satisfied in two different ways, biologists have $2 \times 2 = 4$ metabolic terms for organisms: *chemoheterotroph*, *chemoautotroph*, *photoheterotroph*, and *photoautotroph* (Fig. 6).

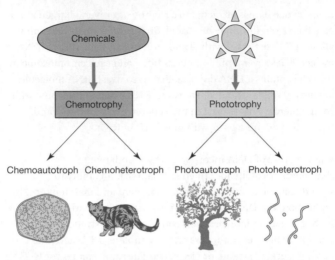

6. The classification scheme for metabolisms in terrestrial life

All organisms (probably even extraterrestrial ones) fall into one or more of these categories. The *troph* suffix means 'to feed' and the two ways in which organisms get energy, from chemical sources or sunlight, give rise to the *chemo-* and *photo-* prefixes. An additional *hetero-* or *auto-* prefix is employed depending upon the method used to acquire carbon. If carbon is obtained by consuming organic carbon compounds (such as sugars), the *hetero-* prefix applies. If an organism converts inorganic carbon (e.g. carbon dioxide) into organic carbon—called 'fixing carbon'—the *auto-* prefix is used. In general, heterotrophs must acquire food to make energy, while autotrophs can fix carbon and make their own energy.

We humans are chemoheterotrophs. You and I consume organic chemicals made by other organisms, such as plants. All animals and fungi, many protists, and most known microbes are chemoheterotrophs. In contrast, a chemoautotroph is a microbe that uses inorganic chemicals such as hydrogen, hydrogen sulphide, iron, or ammonia to make energy, some of which it uses to extract carbon from carbon dioxide. For example, chemoautotrophic microbes live in darkness in deep-sea hydrothermal vents by oxidizing hydrogen sulphide and other substances. (Chemoautotrophs are also called *chemolithotrophs*, from the Greek 'lithos' for stone because the inorganic chemicals they use come from geological sources.) Plants, algae, and some cyanobacteria are all photoautotrophs because they use sunlight for energy and acquire carbon from the air. Photosynthetic bacteria from the *Chloroflexus* genus, which are found in hot springs, are an example of microbes that metabolize as photoheterotrophs using sunlight for energy and acquiring carbon from organic compounds made by other microbes. Chemoautotrophs might inhabit the subsurface of Mars or Europa, while phototrophs might exist in the oceans of habitable exoplanets.

The tree (or web) of life

The history of life on Earth can be deduced from the way that evolution has altered genes. Evolution is the change in inherited characteristics in a population from one generation to the next. Because individuals are genetically variable, in any given environment some will be better adapted and have greater reproductive success than others, which biologists describe as higher *fitness*. In every generation, individuals of lower fitness are lost. This is natural selection. So, over many generations, lineages accumulate genetic adaptations and new species evolve.

In sexual organisms, *species* are groups that cannot interbreed under natural conditions, such as humans and horses. Many bacteria and archaea live in different ecological niches that define separate groups, but the ease with which genes can move between microbes makes it more difficult to designate species of microbe. Consequently, genetics is used. A difference of about 2–3 per cent in the genes that code the RNA contained in ribosomes is enough to separate microbial species.

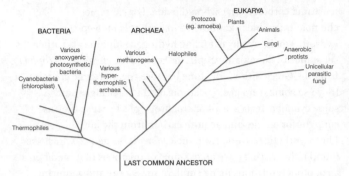

7. The 'tree of life' constructed from ribosomal RNA

In fact, we can assess the relatedness of all life forms by comparing genes. For example, genes indicate that you and the fungus between your toes—which are both eukaryotes—are much closer relatives than an archaeon and a bacterium, despite appearances. From the mid 1960s onwards, various techniques have been developed to assess organisms at the molecular level, including a comparison of the sequences of either amino acids in proteins or nucleotides in RNA and DNA.

Genes define the sequence of amino acids in a protein, so differences in the 'protein sequence' between species indicate disparity or relatedness. For example, 'cytochrome c' is a respiratory protein of 104 amino acids found in many organisms. The sequences are identical in humans and chimps, showing their close relationship. Compared to the human protein, a rhesus monkey has one different amino acid, a dog has thirteen different amino acids, and so on. As species become more distantly related, the protein sequences diverge. With this sort of data, an analyst can draw up a tree, like a family tree, that relates all the species.

In the 1970s, the American microbiologist Carl Woese (1928–2012) used nucleic acids rather than proteins to determine relationships among organisms. He examined RNA in ribosomes (rRNA, for short), with the key insight that this enabled analyses of all life. Protein synthesis, the job of ribosomes, is a core function of any cell. Consequently, the genes that code for rRNA should have mutated slowly over time because most mutations for this process would have been fatal and so failed to accumulate. About 60 per cent of the dry weight of ribosomes is made up of rRNA, and the remainder is protein. Ribosomes in prokaryotes are smaller than in eukaryotes, but otherwise similar in structure and function, allowing their comparison.

Woese isolated a small rRNA and by comparing the differences in the nucleotide sequences, he built a 'tree of life' (Fig. 7). The construction of the evolutionary history of organisms in this way is called *phylogeny*. Woese's tree was a shock because life grouped into the three domains mentioned earlier, unlike the 'five kingdom' paradigm in which archaea were lumped with the bacteria. Furthermore, Woese's tree demoted plants, animals, and fungi to mere twigs at the end of the eukaryotic branch.

Today, while Woese's conception of three domains is widely accepted, studies of other genes show that a tree with vertical descent from one generation to the next is too simple. This is because microbes swap genes willy-nilly (Fig. 4), sometimes to unrelated species, which is called *lateral* (or *horizontal*) *gene transfer*. Thus, microbial branches of the tree of life are more like a 'web of life', criss-crossed by lateral gene transfers.

Since the mid 1990s, DNA sequencing has become more automated. For sequencing a single gene from environmental samples, the process is as follows: isolate DNA from cells → copy (or 'amplify') a DNA gene many times using a procedure called the 'polymerase chain reaction' → obtain the gene sequence → compare with other organisms → produce a phylogenetic tree. Increasingly, whole genomes are sequenced, including the human genome, which contains roughly 21,000 protein-coding genes.

Astonishingly, human protein-encoding genes cover only 1.5 per cent of 3 billion nucleotides. The rest of the sequence is said to be 'non-coding' or 'junk DNA'. These terms are actually misnomers because many parts of non-coding DNA regulate when certain genes are expressed (i.e. give rise to proteins) or code for non-protein products such as rRNAs.

The surprising division between archaea and bacteria revealed by genes is confirmed by biochemical differences. For example, bacterial cell walls contain peptidoglycan, which consists of carbohydrate rods cross-linked by proteins, whereas archaeal cell walls have variable chemistry of protein or carbohydrate, or both. Eukaryote cell walls, for comparison, are made of cellulose in plants, chitin in fungi, and are non-existent in animal cells, which have only a membrane. Furthermore, the cell membrane lipids of bacteria and archaea differ. First, there are dissimilar chemical linkages between the hydrophobic and hydrophilic ends of phospholipids. Second, instead of a lipid bilayer with hydrophobic ends dangling face-to-face in the middle of the membrane, archaea have molecules that connect all the way through. This provides strength and is one reason why some archaea can live in very hot water as hyperthermophiles. There are also ecological distinctions. For example, no archaeon is a pathogen, so you'll never get a disease from archaea whereas you can from various bacteria.

One further use of phylogeny is that the genetic difference between taxa can be related to the time in geological history that they diverged, making a *molecular clock*. In the 1960s, Emile Zuckerkandl and the Nobel Prize-winning chemist Linus Pauling compared proteins for taxa that were known from fossils to have diverged from a common ancestor. They found that the number of amino acid differences is proportional to elapsed time. One interpretation is that most changes are 'neutral mutations' that have no effect on fitness. Similarly, the number of nucleotide substitutions in certain DNA sequences is proportional to elapsed time. Molecular clocks work best with closely related groups of species that are likely to have had similar rates of mutation. Distantly related species have disparate generation times and metabolic rates with variable mutation rates that must be taken into account. To use a molecular clock, a calibration point from the fossil record fixes the date of a particular ancestor in a computer algorithm applied to the molecular data. As we go back very deep into time, there are fewer fossils, so the technique becomes challenging. Nonetheless, molecular clocks indicate that the last common ancestor of animals occurred about 750–800 Ma, which is curious because it predates the oldest animal fossils.

Despite the complications of lateral gene transfers, the tree of life provides information about early life. Thermophiles (microbes that thrive at high temperatures) that grow best at 80–110°C are found near the root of the rRNA tree in the archaea and bacteria (Fig. 7). A reasonable inference is that the last common ancestor lived in a hydrothermal environment. Organisms close to the root are also chemoautotrophs, suggesting that primitive microbes probably gained energy from inorganic compounds. The tree also shows that complex organisms—the plants, fungi, and animals—were late to evolve, consistent with the fossil record. Consequently, if Earth's phylogeny is a guide for astrobiology, the implication is the same as the fossil record: life elsewhere ought to be mostly microbial for much of the history of the host planet.

Life in extreme environments

Apart from being clustered near the last common ancestor, thermophiles were the starting point for research into *extremophiles*. Extremophiles are organisms that thrive under environmental conditions that are extreme from a human perspective. In 1965, Thomas Brock, an American microbiologist, discovered pink filaments of bacteria living at temperatures of 82–88°C in a steamy hot spring in Yellowstone National Park. At that time, no life was known to exist above 73°C, so Brock's discovery spurred an interest in exploring the limits of life.

Although it wasn't anticipated, Brock's efforts also ultimately enabled the explosion in genetics. Brock found a new bacterium, *Thermus aquaticus*, in another hot spring. From this microbe, industrial scientists isolated an enzyme stable at high temperatures that was able to catalyse the polymerase chain reaction (PCR)—the DNA duplication technique that revolutionized biology. Thus, pure science—in this case, what we now call astrobiology—ended up benefiting society unexpectedly. Today, PCR technology is a multibillion-dollar industry. Brock, however, gave away his bacterium and didn't get a penny.

Often, the growth (meaning replication) of extremophiles either requires extreme conditions or is optimal at them. For temperature, the upper limit for eukaryotes is 62°C, compared to 95°C for bacteria and 122°C for archaea. The record holder, the methane-producing archaeon, or methanogen, *Methanopyrus kandleri*, has optimal growth at 98°C.

Identification of various extremophiles (Box 1) indicates that life exists in a much wider range of environments than anyone thought feasible fifty years ago, which opens up possibilities for life beyond Earth. For example, thermophiles might survive deep underground on Mars because they exist deep in the Earth's crust. Another example concerns life in Lake Vostok (250 km long, 50 km wide, and 1.2 km deep), which sits below 4 kilometres of ice in east Antarctica. Ice some 100 metres above the lake is thought to have frozen from lake water. Oddly, it contains traces of the DNA of chemoautotrophic thermophiles. This suggests that beneath the cold lake (at −2°C because of the high pressure), there may be hydrothermal water containing

Box 1 Extremophiles

Acidophiles: require an acidic medium at pH 3 or below for growth; some tolerate a pH of below 0.

Alkaliphiles: require an alkaline medium above pH 9 for optimal growth and some live up to pH 12.

Barophiles (or piezophiles): live optimally at high pressure; some live at over 1,000 times the Earth's atmospheric pressure.

Endoliths: live inside the pore space of a rock (*endo* = within, *lithos* = stone).

Halophiles: require a salty medium, at least a third as salty as seawater (*halo* = salt).

Hypoliths: live under stones (*hypo* = under).

Psychrophiles: grow best below 15°C, while some grow down to –15°C, with reports as low as –35°C for metabolism.

Radioresistant microbes: resist ionizing radiation, such as from radioactive materials.

Thermophiles: thrive at temperatures between 60 and 80°C, while a *hyperthemophile* has optimal growth above 80°C; they can be contrasted with *mesophiles*, such as humans, which live between 15 and 50°C.

Xerophiles: grow with very little available water, and may be halophiles or endoliths (*xeros* = dry).

thermophiles emanating from fractures. The techniques honed for detecting life in lakes such as Vostok can be applied in the search for life elsewhere in the Solar System. Sampling for life in the subsurface ocean of Jupiter's moon, Europa, involves similar challenges.

Chapter 6
Life in the Solar System

Which worlds might be habitable today?

In 2002, while teaching astrobiology, I offered a prize to any student who could guess nine celestial bodies up to the orbit of Pluto that I reckoned might possibly harbour extraterrestrial life (Table 1). No one won. But, with the growth of astrobiology discoveries and online information, someone received the prize in 2010. Today I might add several more bodies, but we'll leave those until the end of this Chapter.

The type of life that I'm considering in Table 1 is simple, comparable to microbes, and the guiding principle concerns liquid water. On Earth, wherever we find liquid water, we find life, whether in bubbling hot springs, drops of brine inside ice, or films of water around minerals deep in the crust.

Some go further and speculate that *weird life*—organisms that don't depend on water—might exist in lakes on Titan, the largest moon of Saturn. Titan's lakes contain liquid hydrocarbons, not water, somewhat like small seas of petroleum. Life that uses hydrocarbon solvent is unknown, and there are some arguments from physical chemistry, which I'll mention later, that such life might be difficult. In contrast, there's no question that liquid water can support life.

Table 1. Nine abodes where life might exist in the Solar System today. The distance to the Sun is in Astronomical Units (AU), where 1 AU is the Earth–Sun distance of about 150 million kilometres or 93 million miles

Body	Type of body and average distance from the Sun	Why it might have life
Mars	planet, 1.5 AU	might have subsurface pockets of liquid water
Ceres	largest asteroid, 2.8 AU	might have a subsurface ocean
Europa, Ganymede, Callisto	large icy moons of Jupiter, 5.2 AU	evidence for subsurface oceans
Enceladus	icy moon of Saturn, 9.8 AU	evidence for a subsurface ocean or sea and presence of organics
Titan	largest moon of Saturn, 9.8 AU	evidence for a subsurface ocean and presence of organics
Triton	largest moon of Neptune, 30.1 AU	might have a subsurface ocean
Pluto	large Kuiper Belt object, 39.3 AU	might have a subsurface ocean

Sunlight and the habitability of the inner planets

Whether the inner planets—Mercury, Venus, Earth, or Mars—have liquid water mostly has to do with temperature and, thus, distance from the Sun (Table 2). 'Location, location, location' doesn't just sell houses but is critical for planetary habitability.

Table 2. The inner planets and factors that affect their current habitability

	Mercury	Venus	Earth	Mars
Average distance to Sun (Astronomical Units)	0.4	0.72	1	1.5
Average surface temperature (°C)	167	462	15	−56
Greenhouse effect (°C)	0	507	33	7
Atmospheric pressure (bar)	0	93	1	0.006
Liquid water?	Too hot	Too hot	Abundant	Mostly too cold
Main gases in the atmosphere	airless	96.5% carbon dioxide, 3.5% nitrogen	78% nitrogen, 21% oxygen	95.3% carbon dioxide, 2.7% nitrogen

Location matters because sunlight spreads into a sphere with a surface area that grows with the square of the planet–Sun distance. *Solar flux* is the wattage (like a light bulb) of sunlight received per square metre. At the Earth's orbital distance of 1 Astronomical Unit (AU), sunshine provides 1,366 Watts over every square metre, the equivalent of almost fourteen 100-Watt light bulbs. At the 5 AU distance of Jupiter, the solar flux is a factor of 25 times smaller because the same energy spreads over a sphere that has $5 \times 5 = 25$ times the area. For Mars, at 1.5 AU, the solar flux is 2.25 ($= 1.5 \times 1.5$) times smaller than for Earth. In contrast, Venus, at 0.72 AU, receives solar flux almost twice as big as that of the Earth.

Mercury, at only 0.4 AU from the Sun, is lifeless. It's the smallest of the eight planets (about two-fifths the diameter of Earth) and has no liquid water and probably never did. Today, Mercury's barren surface sometimes reaches 430°C. If Mercury once had an atmosphere, it would have burned off when Mercury formed because of low gravity and intense sunlight.

The distance from the Sun explains why Venus and Mars are hostile to life, although that's not the whole story. The roughly 460°C surface of Venus is even hotter than Mercury gets, while Mars is an icy desert with an average temperature of −56°C. Venus is comparable in size to the Earth, but Mars is smaller—around half the diameter and one-ninth the mass of Earth. In fact, Mars's small size led to its poor habitability today, as we'll see later.

Although solar flux is one factor, the greenhouse effect on Mars and Venus also determines surface temperatures. The Martian atmosphere exerts only 0.006 bar surface pressure, compared to Earth's 1 bar. This wispy air is an arid mix of 95.3 per cent carbon dioxide, 2.7 per cent nitrogen, and minor gases. Recall that Earth's greenhouse effect is 33°C. Because of atmospheric thinness and dryness, Mars's greenhouse effect is only 7°C, which leaves the planet frozen. The two main gases in Venus's atmosphere, 96.5 per cent carbon dioxide and 3.5 per cent nitrogen, have similar proportions to those on Mars. But in stark contrast, Venus's atmosphere is massively thick and has a huge pressure of 93 bar at an average ground elevation. As a result, Venus's greenhouse effect is a whopping 507°C, i.e. 500°C bigger than on Mars (Table 2). The upshot is that neither planet's surface supports liquid water. In the thin air on Mars, a puddle would boil (or rapidly evaporate) and freeze at the same time, and water generally exists only as ice or vapour. Meanwhile on the scorching surface of Venus, liquid water is impossible.

Without liquid water, Venus is considered by most astrobiologists to be lifeless. A few speculate that acidophile microbes might live in its clouds of sulphuric acid particles. However, I doubt it. Apart from the lack of available water, atmospheric turbulence would pull microbes down into Venus's inferno or up to fatally desiccating heights.

Was Venus inhabited in the past?

Venus is still astrobiologically interesting because it may once have had oceans and life. It should have begun with plenty of water because the amounts of other volatiles are similar to Earth's. (Volatiles are substances that can become gases at prevailing planetary temperatures.) For example, if you took all of the Earth's carbonate rocks and turned them into carbon dioxide (CO_2), the Earth would have about ninety atmospheres' worth of CO_2, like Venus. If you extracted all the nitrogen in minerals on the Earth and added it to the Earth's atmosphere, you would get almost three atmospheres' worth of nitrogen gas, which again is similar to that in Venus's atmosphere. Because accretion of hydrated asteroids was the ultimate source of carbon and nitrogen on Venus and Earth, it's reasonable to infer that Venus also gained lots of water from them, as the Earth did.

Unfortunately Venus was doomed by its proximity to the Sun because of a *runaway greenhouse effect*. When baked by intense sunlight, the evaporation of water can make an atmosphere so moist that it becomes completely opaque to the infrared radiation emitted from the planet's surface. At this point, there's a limit to the cooling of the planet by emission of infrared radiation to space, which is set by the properties of water and a planet's gravity. Recent calculations suggest that this *runaway limit* is about 282 Watts per square metre for Earth and a few Watts less for Venus.

To understand the runaway limit, consider turning Earth into Venus. As mentioned earlier, the solar flux at the Earth's orbit is 1,366 Watts per square metre but the Earth reflects 30 per cent of this and so absorbs only 70 per cent. Then an additional reduction of 50 per cent comes from having only half of the Earth in daylight, and a further 50 per cent decrease accounts for glancing sunbeams on Earth's curved surface. Putting all these factors together, the Earth absorbs a net $(0.7 \times 0.5 \times 0.5) \times 1,366 = 240$ Watts per square metre of sunlight. When stable, the Earth emits the same amount of energy into space in the infrared and so keeps a constant global average temperature. But imagine if the Earth were moved to the orbit of Venus, where the absorbed sunlight per square metre would double from 240 to 480 Watts. The ocean would become like a hot bath and the steamy atmosphere would reach the runaway limit where only 282 Watts per square metre can radiate into space. With more energy per square metre coming in (480 Watts) than going out (282 Watts), the Earth would simply get hotter and hotter. The entire ocean would evaporate and surface rocks would melt. At that point, near-infrared light from a searing upper atmosphere would shine through to space, so that the incoming sunlight and outgoing radiation would come back into balance and the surface temperature would plateau around 1,200°C. This nasty process is what we think happened to Venus. Any Venusians would have been toast.

The runaway would also cause Venus's ocean to vanish. In the upper runaway atmosphere, ultraviolet light would split water vapour into hydrogen and oxygen. Hydrogen is so light that it would escape into space, dragging along some oxygen, while any oxygen left behind would oxidize hot rocks below. Eventually rocks would solidify, but by this time the atmosphere would be loaded with carbon dioxide and nitrogen released from the hot surface. The end result would be today's hellish Venus.

Before the runaway, life might have thrived in Venus's oceans. Recall that the early Sun was less luminous. So, a few hundred million years might have passed before the runaway commenced. Unfortunately, it might be tricky to prove if life ever existed. In the 1990s, radar on NASA's *Magellan* spacecraft mapped the Venusian surface. Older surfaces have more craters, so the density of impact craters was used to estimate the age of the surface as a fairly uniform 600–800 Ma. Venus's surface appears to have been repaved by lava all at once. A possible reason is that whereas Earth's internal heat is released continuously by creating new seafloor, on Venus internal heat might periodically build up until the interior becomes so hot that lava erupts everywhere. Venus might behave like this because unlike Earth it doesn't have water to lubricate plate tectonics.

Resurfacing would destroy any fossils. However, subtle geochemical traces might remain. To investigate if Venus had life, the best option would be to collect samples of Venusian rocks and return them to Earth in a future space mission.

Watery Mars: an abode for life?

Many have imagined Mars, unlike Venus, as a potential abode for life. Before the Space Age, basic parameters were known, such as Mars's 24.66 hour day, a year of 1.9 Earth years, and gravity that is 40 per cent of Earth's. Not much about the surface was certain until the first successful space mission, the flyby of NASA's *Mariner 4* in 1965, revealed a heavily cratered surface like the Moon. This put a damper on hopes for life. Then, in the early

1970s, NASA's *Mariner 9* orbiter photographed dried-up river valleys and extinct volcanoes, suggesting that Mars had once been quite Earth-like after all. But soon the pendulum swung back in the other direction. The *Viking* mission, consisting of two identical orbiters and landers, reached Mars in 1976, and failed to find life, as we discuss later.

After a hiatus in the 1980s, exploration was revived. *Mars Global Surveyor*, which orbited from 1997 to 2006, mapped Mars and imaged sedimentary layers that implied many geologic cycles of erosion and deposition. In the early 21st century, the *Mars Odyssey* and *Mars Express* orbiters, along with *Mars Reconnaissance Orbiter*, discovered areas of clay minerals and salts, which may have formed in liquid water. Twin *Mars Exploration Rovers* landed in 2004, and found fossilized ripples from past liquid water and sedimentary rocks, while in 2008, NASA's *Phoenix Lander* dug up subsurface ice in a polar region and measured soluble salts in the soil. Finally, in 2012, *Curiosity Rover* trundled towards a 5-kilometre-high mountain of sedimentary beds within a 150-kilometre-diameter crater named Gale. It found mudstones deposited from water that contain the SPONCH elements (Chapter 1) needed for life.

Nowadays, astrobiological interest in Mars concerns either biological traces from billions of years ago or microbial-like life that miserably endures underground. Pre-Space Age hopes of an Earth-like Mars today have been replaced with the reality of a cold, windswept, global desert with dust storms, daily dust devils, and no rainfall.

Mars's current surface is hostile to life for three reasons. First, while ice exists for sure, no liquid water has been unequivocally identified. The polar caps are water ice, topped with carbon dioxide 'dry ice' that grows when about 30 per cent of the atmosphere freezes at the winter pole. Also, above mid latitudes, ice-cemented soil or *permafrost* lies just beneath the surface. In the tropics, afternoon temperatures in the top centimetre of soil rise above freezing, but there, ice turns to vapour before melting temperatures are reached. A second problem is no ozone layer, allowing harmful ultraviolet sunlight to reach the surface. Third, chemical reactions in the atmosphere make hydrogen peroxide—the same chemical used in hair bleach. Hydrogen peroxide molecules settle to the surface, where they can destroy organics.

While the surface is unpromising, geothermal heat underground might allow liquid water and life to exist. In fact, from 2004, reports of atmospheric methane averaging ten parts per billion by volume led to excitement that subterranean methanogens might be present. Sunlight reflected from Mars gathered by telescopes and *Mars Express* appeared to show absorption by atmospheric methane. However, the methane signal was barely distinguishable and some sceptical scientists (including me) doubted whether methane was really present. Subsequently, the *Curiosity Rover* has failed to detect methane down to levels of one part per billion.

In discussing past life, astrobiologists refer to the Martian geological timescale, which is divided into aeons called the *Pre-Noachian* (before 4.1 Ga), *Noachian* (4.1 to about 3.7 Ga), *Hesperian* (3.7 to 3.0 Ga), and *Amazonian* (since 3.0 Ga). Surfaces on Mars are placed into each aeon according to impact craters. Older surfaces have accumulated bigger and more numerous craters. In fact, the dates bounding each aeon actually come from the Moon. The ages of rocks brought back by the Apollo astronauts are known from radioisotopes and these correlate lunar cratering densities with time. Astronomical calculations that account for more impactors on Mars than the Moon allow the lunar correlation to be extended to Mars.

Evidence that liquid water used to be present suggests that Mars was once more habitable than today. Images show *fluvial* (stream-related) features in the landscape, including gullies, dried-up river valleys, deltas, and enormous channels. Also, the soil and rocks contain minerals that form in the presence of liquid water.

Gullies are incisions of tens to hundreds of metres length on the walls of craters and mesas between 30 and 70° latitudes in both hemispheres. Because gullies lack superimposed craters and sometimes flow over sand dunes, they must be very recent. Initially, they were thought to form when ice melted. However, images show them forming when carbon dioxide frost vapourizes, presumably releasing a dry flow of soil and rocks.

Far more ancient features are *valley networks*, which are dried-up river like depressions that spread out in tree-like branches with tributaries (Fig. 8). Most incise heavily cratered Noachian terrain and they're 1–4 km wide and 50–300 m in depth. The density of tributaries is far less than for most terrestrial rivers, but sometimes enough to infer the drainage of rain or meltwaters. In other cases, valleys with stubby tributaries were probably formed by *sapping*, when underground (melt)water caused erosion and collapse of the overlying ground. A few valleys end in deltas.

The Noachian landscape consists of craters with degraded rims and shallow floors. What caused this erosion is unclear. Valley networks are incised on top and so weren't responsible. Once the Hesperian started at 3.7 Ga, erosion rates dropped dramatically and valley networks became rare.

Towards the end of the Hesperian (around 3 Ga), *outflow channels* appeared (Fig. 8). Channels form from fluid flow confined between banks, lacking tributaries and emerging from a single source, unlike river valleys. The outflow channels are huge: 10–400 km width, up to 1,000 kilometres or so long, and up to several kilometres deep. Most begin in *chaotic terrain* where the ground collapsed, sometimes in canyons or chasms where there

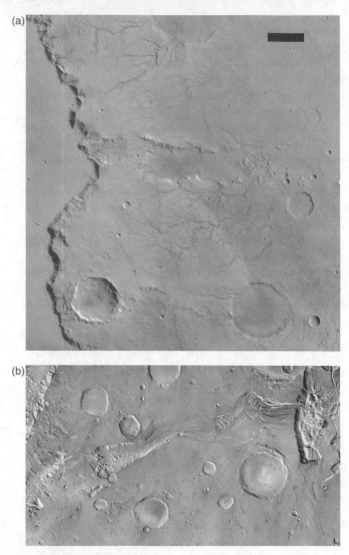

8. a) Valley networks on Mars. On the left is the eroded rim of
456-km-diameter Huygens Crater. The image is centred at 14°S, 61°E,
north upwards. Scale bar = 20 km; b) Outflow channel Ravi Vallis
(0.5°S, 318°E) about 205 km long

are mountain-sized heaps of sulphate salts, which might be evaporation residues from salty water.

The leading explanation for outflow channels is that they formed from floodwaters when underground ice melted or aquifers burst. However, they require 10–100 times more flow than the best-known terrestrial analogue: flood-carved land across eastern Washington State, USA, which early settlers called *scablands*. The scablands formed at the end of the last ice age when ice damming of a large lake periodically ruptured.

It's not easy to explain the water needed to erode Mars's outflow channels. Estimates suggest the equivalent of a global ocean several hundred metres deep, which is far more water than exists as ice today. Since the channels flow to the northern lowlands, some scientists speculate that an ocean formed there. In contrast, a minority thinks that the outflow channels were not carved by water but by lavas. Chemistry suggests that Martian lavas should have been runny, turbulent, and erosive.

Apart from ancient valley networks and outflow channels (if water eroded), minerals provide other evidence for a wetter early Mars. Everyone knows how lettering on old gravestones disappears because of chemical reactions with water. Such reactions change or dissolve minerals in *chemical weathering*. Much of the Martian surface, like that on Venus and the Earth's seafloor, is made of basalt, a dark-coloured igneous rock rich in iron and magnesium silicate minerals. When basalt is chemically weathered, *alteration minerals* are produced, such as clays. So the presence of alteration minerals means that liquid water was present, sometimes with a specific pH. For example, alkaline waters tend to produce clay minerals from basalt.

Hydrous (water-containing) alteration minerals have been identified from analysis of infrared radiation emitted and reflected by Mars's surface. But only about 3 per cent of Noachian surfaces have hydrous clays and carbonates. Sulphate minerals dominate late Noachian or Hesperian areas, while reddish, dry iron oxides are common on younger Amazonian surfaces. This pattern might imply three environmental epochs. In the first epoch, alkaline or neutral pH waters weathered basalt and made clays. During the second, sulphuric acid was derived from volcanic sulphur gases and made the sulphates. The third epoch continues today with a cold, dry environment and rust-coloured surfaces.

One of the twin *Mars Exploration Rovers*, named *Opportunity*, actually landed near Noachian sulphates. Millimetre-sized spheres of the iron oxide hematite (Fe_2O_3) were embedded in sulphate layers, like blueberries in a muffin. The hematite precipitated from minerals carried in water percolating underground about 3.7 billion years ago. Also, ankle-deep water appears to have ponded on the surface, leaving behind ripples in the sediment. The other Rover, named *Spirit*, found evidence of ancient hot springs on the opposite side of the planet.

The early atmosphere and climate of Mars

Evidence of liquid water leaves us wondering whether the climate on early Mars was warm and wet. Scientists disagree and fall into two camps. One group argues that Mars had a warm, wet climate for tens or hundreds of millions of years. The other maintains that transient melting of ice in a cold climate could account for what is seen.

The first scenario is obviously more favourable for life. But unfortunately, no one has yet explained how early Mars was kept warm for millions of years. Some 3.7 billion years ago, the Sun was 25 per cent fainter. A greenhouse effect of about 80°C would have been needed to keep early Mars just above freezing, compared to Earth's modern 33°C greenhouse. It's generally thought that any atmospheric hydrogen should have escaped into space rapidly when Mars formed, leaving the atmosphere oxidized and full of carbon dioxide and nitrogen. Mars couldn't ever have had an extremely thick Venus-like atmosphere because carbon dioxide would condense into ice and form clouds at Mars's distance from the Sun. A thick carbon dioxide atmosphere is also good at scattering sunlight back to space, which cools the surface. So carbon dioxide can't provide an early warm climate. An alternative suggestion is that volcanic sulphur dioxide was the key greenhouse gas. However, sulphur dioxide dissolves in rain and would be flushed from the atmosphere if Mars became wet. Also, atmospheric reactions at high altitudes make a fine suspension of sulphate particles from sulphur dioxide, which reflects sunlight and cools the planet. Such cooling happened on Earth during 1991–3 as a result of volcanic emissions from Mt Pinatubo in the Philippines.

The other camp proposes many mechanisms that allow liquid water in a cold climate. They note that impacts would have vaporized ice into steam, which, in turn, would have produced rainfall that eroded river valleys. Also, erosion might have been produced by local snowmelt as a response to past fortuitous combinations of Mars's axial tilt and orbital shape. These characteristics are perturbed over time by the gravitational influence of other planets. The tilt of a planet's axis with respect to the planet's orbital plane, e.g. 23.5° for Earth, causes the seasons because one hemisphere gets more sunlight at one point in the orbit than the other. Unlike Earth, Mars has no big moon to stabilize its axis (its two tiny moons have negligible effect), so Mars's tilt has varied between 0° and 80° over the last 4.5 billion years. At high tilts, summertime polar ice faces the Sun and vapourizes. Air currents transport the vapour to the cold tropics where it snows. Sunlight during other seasons or at lower tilts could then produce meltwaters and fluvial erosion. Moderate tilts in the last few million years might explain relict midlatitude patches of dust that were once ice-cemented. Finally, salty water on Mars can remain liquid far below 0°C. On Earth, we spread sodium chloride on icy roads in winter because it melts ice down to –21°C. Another salt, perchlorate, which was detected in Martian soil by the *Phoenix* lander in magnesium or calcium form, can depress the freezing point of water below –60°C.

Whatever the truth about early Mars, Mars's small size ultimately spoiled its habitability. Because small objects cool faster than large ones, internal heat was lost rapidly, so that widespread volcanism ceased long ago. Without volcanism, Mars couldn't recycle carbon dioxide, leaving the gas to be converted into carbonates, which are present but not in the abundance expected if a thick carbon dioxide atmosphere had all been transformed. So, in addition, part of the early Martian atmosphere was probably blasted away to space by large comet and asteroid impacts in so-called *impact erosion*. The atmosphere was vulnerable because of Mars's low gravity. Along with more gradual escape of gases to space later, Mars was ultimately left with its thin atmosphere.

Looking for life on Mars

We still don't know if Mars has life or ever had life, despite attempts to find it. The *Viking* landers tried to detect life in 1976, and, since the 1990s, scientists have looked for signs of past life in Martian meteorites.

Each *Viking* lander had three experiments to detect metabolism and a fourth to find organic molecules. The first, the *carbon assimilation experiment*, examined if Martian microbes obtained carbon from air. Martian soil (sometimes mixed with water) was exposed to carbon dioxide (CO_2) and carbon monoxide (CO) gases brought from Earth, with carbon-14, a radioactive isotope, in each gas. Afterwards, the soil was found to have incorporated carbon-14. When a 'control experiment' was done where the soil was first sterilized at 160°C, the soil still took up carbon-14, which suggested that inorganic chemistry was responsible, not Martian microbes. A second test was the *gas exchange* experiment, which monitored Martian soil and a solution of organics brought from

Earth to see if gases were generated from metabolism. Strangely, O_2 was released, but it was also released from sterilized soil. Evidently, chemicals in the soil decomposed into O_2 with water or heat. The third investigation, the *labelled release experiment*, added organics containing carbon-14 to soil. If Martians metabolized the organics, they would give off carbon-14-containing CO_2. Radioactive gas was emitted, while sterilized soil released no radioactive gas. At face value, this was a positive detection of life. But most scientists think that soil oxidants reacted with organics to make CO_2 and were inactivated by heat. A reason to prefer this explanation came from the fourth experiment, in which a sort of electronic nose, called a *gas chromatograph mass spectrometer*, identified molecules wafting off heated soil. It found no organic material in the soil to a detection limit of a few parts per billion.

Overall, the moral is that before looking for extraterrestrials, you need to understand the inorganic chemistry of the environment to avoid false positives. In fact, Harold 'Chuck' Klein (1921–2001), who led the *Viking* biology experiments, told me that he wanted to do this when the *Viking* mission was conceived, but managers at NASA who controlled finances instead insisted on looking for life directly.

While the Viking results were being pondered in the 1980s, a remarkable discovery was made: there were already rocks from Mars on Earth—Martian meteorites! Impacts knock rocks off Mars and some of these fall on the Earth. In fact, about 50 kg lands every year, mostly in the ocean. Gas sealed within some meteorites during minor melting associated with impact ejection proves a Martian origin because it matches the atmosphere measured by the *Viking* landers. Other Martian meteorites without trapped gas have a triple oxygen isotope (^{16}O, ^{17}O, ^{18}O) composition in their silicate minerals that is unique to all Martian meteorites and identifies them like a fingerprint. By 2013, sixty-seven Mars meteorites were known but the list keeps growing.

In 1996, possible signs of past life in ALH84001 were reported. This Martian meteorite was the first one (the '001') collected in the Allan Hills, Antarctica (the 'ALH'), in a 1984 expedition (the '84'). ALH84001 crystallized as an igneous rock on Mars at 4.1 Ga. Inside the rock are some carbonate globules about 0.1 mm across that formed at 4.0–3.9 Ga, and within them are four possible traces of life: alleged microfossils; carbonates said to be precipitated by microbes; traces of organics called polycyclic aromatic hydrocarbons (PAHs), which are made of hexagonal rings of carbon atoms; and crystals of the mineral magnetite (Fe_3O_4) said to be similar to those within certain bacteria. On Earth, magnetotactic bacteria make magnetite crystals inside their cells with shapes that evolution has honed into magnets. The bacteria use magnetite compasses to move along the up–down component of the Earth's magnetic field in order to find a boundary between a lower oxygen-poor zone below and an upper oxygenated zone, which optimizes their metabolism.

In subsequent years, research has cast doubt on all four claims. The alleged microfossils are rod-shaped structures that merely *look like* microbes. Scientists have since found inorganic mineral surfaces with similar shapes. Also, the structures in ALH84001 are about ten times smaller than terrestrial microbes and probably beyond the minimum volume needed for essential biochemistry. An evaporating fluid can produce the carbonates in the globules and so there's little reason to invoke biogenic salts, proposed as the second feature. Regarding the third line of evidence, analyses showed that most PAHs got into the meteorite while it was sitting in Antarctica for 13,000 years. PAHs are ubiquitous atmospheric pollutants produced from burnt organic material. Meteorites are dark and absorb sunlight well, so they can sit in puddles in Antarctic ice during summertime, allowing water and chemicals to infiltrate cracks. If there are Martian PAHs in the interior of ALH8401, these could have been made inorganically. The rock was heated by impacts that occurred before the ejection impact. Heating should have released carbon-bearing gas from the carbonates, which can react with water to make organic matter without life. The fourth argument concerned magnetite. It turns out that only a tiny fraction of magnetites in ALH84001 have biogenic-like shapes. Some scientists argue that magnetite of many shapes, including bacteria-like ones, formed during heat shocks prior to ejection when iron carbonate decomposed into magnetite.

The controversy about ALH84001 shows how difficult it is to prove that microscopic life exists in old rocks. But since most terrestrial rocks lack fossils, absence of life in ALH84001 doesn't mean that life on Mars didn't exist. We need to keep looking.

Ceres: a serious candidate for habitability?

Beyond Mars and within Jupiter's orbit is an asteroid belt of millions of small rocky bodies, including the largest, Ceres, some 950 km in diameter, at 2.8 AU from the Sun. Ceres's surface contains clay minerals and water ice. As on Mars, clays suggest past liquid water. A leading model for Ceres's internal structure is a rocky core that's surrounded by a 100-km-thick shell of ice. Just above the core, there may be a subsurface ocean. This could be very salty, allowing its persistence at sub-zero temperatures, and perhaps it once flowed to the surface.

Could life exist in hydrothermal vents at the bottom of Ceres's ocean? Ceres is so small that there's probably little internal heat available today, so any biomass would be extremely meagre. But perhaps in the past, Ceres was more habitable. NASA's *Dawn* mission will arrive at Ceres in 2015. Spectra of infrared emission and reflected sunlight should give us more detail about the surface composition and habitability.

The icy Galilean moons of Jupiter

Beyond the asteroid belt is Jupiter, which is probably not a realistic target for astrobiology for reasons given in Chapter 5. But

9. The Galilean moons of Jupiter: Io, Europa, Ganymede, and Callisto

Jupiter has sixty-seven moons and three might be habitable.
Galileo discovered the four largest moons in 1610, which are Io,
Europa, Ganymede, and Callisto, going outwards from Jupiter
(Fig. 9). Io and Europa are similar in size to the Moon, while
Ganymede and Callisto have proportions comparable to Mercury.
The outer three satellites have rocky interiors covered with ice,
while Io's exterior is just rocky. Io, Europa, and Ganymede
probably also have iron cores.

Io is the most volcanic body in the Solar System, with volcanoes
spewing sulphur dioxide, which freezes onto Io's surface. But Io is
devoid of liquid water and surely lifeless. Io is so volcanic because
gravitational forces from Jupiter vary as Io goes round in an
elliptical orbit, squeezing Io like a stress ball. This *tidal heating* of
Io is caused by friction when solid material moves up and down,
similar to the way that water sloshes in Earth's oceanic tides. Io's
orbit is forced to be elliptical because of periodic gravitational
prodding from Europa and Ganymede. The times that Ganymede,
Europa, and Io take to make orbits have a ratio of 4:2:1, which
causes the moons to line up cyclically and nudge each other
gravitationally. This relationship is the called the *Laplace
resonance*, after Pierre-Simon Laplace, the French mathematician
mentioned in Chapter 2.

Europa also has an orbit that's forced to be elliptical and so it has
tidal heating too. The heating is smaller than that of Io because
Europa is further from Jupiter. The average temperature of

Europa's icy surface is −155°C, but tidal heat can maintain liquid water deep under the ice. Impact cratering densities suggest a surface age of only 20–180 Ma consistent with slush from below repaving the surface. Images show *chaotic terrain*, i.e. disrupted surfaces with blocks that have shifted, which suggest that subsurface ice might convect. A global network of stripes and ridges includes bands where the surface has ripped apart and ice appears to have been extruded. Magnesium sulphate salt is present on the surface, perhaps originating from waters within, while unknown materials cause reddish streaks.

The key evidence for Europa's subsurface ocean comes from magnetic measurements by NASA's *Galileo* spacecraft, which orbited Jupiter from 1995 to 2003. Unlike Earth or Jupiter, Europa's interior doesn't create its own, *intrinsic* magnetic field. But in passing through Jupiter's large magnetic field, electrical currents are induced inside Europa. In turn, these currents generate a weak and varying *induced magnetic field*. The strength of this field requires that an electrically conducting fluid exist within 200 km of Europa's surface. The most likely explanation is a salty ocean up to twice the volume of Earth's ocean. The largest craters suggest that the ice cover above the ocean is at least 25 km thick. However, some terrain that pokes up might be produced by upwelling ice that melts a few kilometres below the surface and then refreezes. So shallow lens-shaped lakes might exist.

Besides liquid water, the possibility of life depends on energy sources, interfaces, and the availability of SPONCH elements. Europa probably has a good supply of biogenic elements if it was made of material similar to carbon-rich meteorites or comets. Moreover, heat-releasing reactions of water and a rock seafloor could supply energy as well as nutrients. *Radiogenic heat*, which is produced by the decay of radioactive isotopes within the rock (such as potassium, uranium, and thorium), might produce seafloor vents that supply carbon dioxide and hydrogen for chemoautotrophs.

The availability of oxygen in the ocean would also permit biological oxidation of iron and hydrogen on the seafloor. There are two small sources of oxygen. Charged particles trapped in Jupiter's magnetic field slam into Europa's icy surface and break water molecules, releasing oxygen. If the ice churns, some of the oxygen could be carried down to the ocean. Other oxygen comes from water molecules split by radiation from radioactive elements. Estimates of Europa's biomass vary from around a thousand to a billion times less than on Earth. Some optimists even speculate that there might be enough oxygen for animal-like creatures!

Measurements of induced magnetic fields on Ganymede and Callisto also imply subsurface oceans but deeper, below roughly 200 km and 300 km depth, respectively. Ganymede's average surface age is about 0.5 Ga, so its ocean has probably not gushed to the surface recently. Because Ganymede is Jupiter's largest moon, its interior supports high-pressure forms of ice that would lie beneath the ocean. Lack of a rock–water interface might be less favourable for life than on Europa. The lower tidal heating on Ganymede also implies a smaller biomass.

Inference of a subsurface ocean on Callisto is surprising because Callisto lacks tidal heating. The warmth must be radiogenic and the ocean might be loaded with salts that lower the freezing point. Callisto's surface age is 4 Ga, so the prospect of finding evidence for oceanic life there might be slim. Also, with less energy available, any biomass should be even smaller than that on Ganymede.

Enceladus and Titan: icy moons of Saturn

Saturn and its sixty-two moons lie at about twice the distance of Jupiter from the Sun. The largest moons, discovered before the Space Age, are (going outwards): Mimas, Enceladus, Tethys, Dione, Rhea, Titan, Hyperion, Iapetus, and Phoebe.

Enceladus is special because it has active geology. It's Saturn's sixth largest moon, but its 500 km diameter could almost fit within Great Britain. Enceladus orbits Saturn twice for every orbit of Dione. As a result, periodic gravitational nudges from Dione force Enceladus's orbit to be elliptical, resulting in varying gravitational forces from Saturn that flex and heat Enceladus. For reasons that are not fully understood, the heating is concentrated under Enceladus's south pole. There, icy particles and gas spray out of parallel fractures dubbed *tiger stripes*, which are about 130 km long, 2 km wide, and flanked by ridges. The jets contain traces of methane, ammonia, and organic compounds, along with salt. In fact, the jets supply particles to the 'E-ring' of Saturn within which Enceladus orbits.

Enceladus has a rocky core, an icy shell, and an underground sea beneath the area of the jets. The jackpot combination of organic molecules, energy, and liquid water implies that life might exist inside Enceladus. Such life could be chemoautotrophic, feeding off hydrogen produced by water–rock reactions or hydrogen and oxygen from water that's split by radioactivity. If life does exist there, methane or organic matter in the jets could be biological.

Titan, the largest moon of Saturn, is also of astrobiological interest. It's similar in size to Ganymede and Mercury and is the only satellite in the Solar System to have a thick atmosphere. In fact, the air pressure at Titan's surface is 1.5 bar, which is 50 per cent larger than that at the Earth's surface. The atmosphere of 95 per cent nitrogen and 5 per cent methane provides a 10°C greenhouse effect, but the sunlight at Saturn's distance is one hundred times less intense than at Earth, so Titan's surface is incredibly cold at –179°C.

Titan's atmosphere contains a smoggy haze of hydrocarbons. At high altitude, ultraviolet sunlight breaks up methane molecules (CH_4) and subsequent reactions build up hydrocarbons including ethane (C_2H_6), acetylene (C_2H_2), propane (C_3H_8), benzene (C_6H_6), and reddish-brown particles containing polyaromatic hydrocarbons. The particles are called *tholins* from the Greek *tholos* for 'muddy.'

The products derived from methane sediment out of the atmosphere. In fact, about 20 per cent of Titan's surface is covered in tropical dunes that are made of sand-sized particles, which are at least coated with organics if not made of them. At the poles, over 400 lakes are mixtures of liquid propane, ethane, and methane, with extraordinary beaches made of particles of benzene and acetylene.

Methane on Titan behaves like water on Earth and forms clouds, rain, and rivers. Most of our knowledge about Titan comes from the *Cassini–Huygens* mission, which arrived in 2004. *Huygens* landed on Titan in 2005, while *Cassini* is a Saturn orbiter. Near the landing site of *Huygens*, channels were eroded by liquid hydrocarbons (Fig. 10). At the landing site itself, cobbles (presumably made of water or carbon dioxide ice) have been rounded as if by transport in rivers of liquid methane. Methane rain should be very infrequent, but when it rains, it pours.

Because methane is destroyed by sunlight, Titan's atmosphere would run out of methane in about 30–100 million years if it were not resupplied. How methane is replenished is a mystery. The leading hypothesis is that methane leaks out of Titan's interior, which suggests geological movement inside Titan.

Titan flexes too much to be entirely solid, which is evidence that Titan has a subsurface ocean. The eccentricity of Titan's orbit ensures that Titan is squeezed by Saturn's gravity. The ocean exists below an icy crust of less than 100 km thickness. But as with Ganymede, the seafloor should be dense ice rather than rock because Titan's mass allows such high-pressure forms of ice.

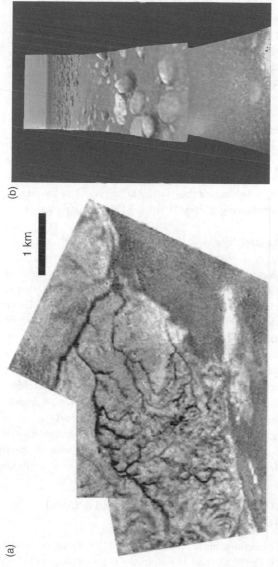

10. a) A network of channels that appear to flow into a plain near the Huygens landing site; b) Image of the surface at the Huygens landing site. Stones in the foreground are 10–15 cm in size and sit on a darker, fine-grained substrate

1 km

(a)

(b)

105

Two types of life might exist on Titan: Earth-like life in the subsurface ocean; and weird life in the hydrocarbon lakes. Regarding the first possibility, when Titan formed, the heat released from the capture of smaller bodies should have created liquid water on the surface temporarily. Perhaps water-based life evolved and survived as an ocean retreated underground.

Weird life would be limited in biochemical complexity because oxygen, which is scarce, is required for sugars, amino acids, and nucleotides. Titan captures some oxygen-containing molecules from space, but the supply is tiny. Hypothetical life in cold hydrocarbon solvents might use acetylene for energy. On Earth, acetylene burns with oxygen in welding torches. But Titanians could metabolize acetylene with hydrogen from Titan's air. The chemical reaction,

$$C_2H_2(\text{acetylene}) + 3H_2(\text{hydrogen}) = 2CH_4(\text{methane})$$

produces energy, but the idea is speculative. The main problem for weird life on Titan, I think, is that liquid hydrocarbons are poor at dissolving large molecules, which are essential for genomes. Larger molecules are less soluble than smaller ones in liquid hydrocarbons. Furthermore, solubility declines at lower temperatures.

Some exoplanets might have Titan-like liquid hydrocarbons. The most common stars are red dwarfs, which are smaller and colder than the Sun. Exoplanets near 1 AU distance around such dwarfs would have surface temperatures in the range of liquid hydrocarbons, not liquid water. Thus, if we were to discover weird life on Titan, we might imagine a universe teeming with life utterly different from our own. Ironically, we would then be the weird life.

Triton: a captured Kuiper Belt object around Neptune

At and beyond the orbit of Neptune at 30 AU, objects are covered in nitrogen ice (N_2). Triton, the largest of thirteen moons of

Neptune, and Pluto are both really *Kuiper Belt objects* (KBOs). The *Kuiper Belt* is a region of icy bodies within the plane of the Solar System at 30–50 AU left over from Solar System formation. There are also KBOs scattered out to 1,000 AU. Triton was captured by Neptune because two features of its orbit suggest so: Triton orbits in the opposite direction to Neptune's rotation, and its orbital plane is tipped up 157° with respect to Neptune's equator.

Triton's surface is tremendously cold, around −235°C, because it reflects 85 per cent of sunlight. An extremely thin nitrogen atmosphere exerts 20 millionths of a bar surface pressure. This 'air' is simply the nitrogen vapour that will sit over nitrogen ice at Triton's prevailing temperature. There's also a little methane (about 0.03 per cent of the nitrogen concentration), which is destroyed by ultraviolet sunlight, creating a thin smog of hydrocarbon particles. If methane destruction has been operating at the same rate for 4.5 billion years, about a metre's depth of organic material should have accumulated. However, mobile frosts would cover this up. A reddish tint to some of the ice might be the organics. In fact, sunlight sometimes vaporizes nitrogen ice into geyser-like plumes, which carry dark particles that are perhaps organic. Apart from nitrogen ice, water and carbon dioxide ice make up some of the surface.

Triton is geologically active because crater counts suggest that resurfacing is only 10 Ma and there are icy structures like vents, fissures, and lavas. We call these *cryovolcanic*, meaning a form of volcanism in which slushy ice comprises the equivalent of lava and magma. Triton's density suggests a large rocky core that supplies sufficient radiogenic heat for cryovolcanism and potentially a subsurface ammonia-water ocean.

Early Triton may have been more habitable. Just after capture, Triton's orbit would have been highly elliptical, producing huge tidal heating from Neptune that should have melted an extensive subsurface ocean. Tides tend to circularize orbits and today Triton's orbit is close to circular. Thus, after capture, the ice shell gradually thickened as tidal heating declined. Today, the ocean might be below hundreds of kilometres of ice, and would be underlain by high-pressure forms of ice. But although oceanic life might be deeply hidden, the presence of cryovolcanism suggests that if we sample organic matter on the surface, there's a chance of finding traces of life.

Does Pluto have an underworld ocean?

Pluto, named after the god of the underworld, has a surface mainly of nitrogen ice, and a thin nitrogen atmosphere of 8–15 microbar surface pressure. There's also a little atmospheric methane that sunlight destroys just as on Triton and Titan. Again, this produces hydrocarbon particles, which settle out, possibly accounting for dark areas on Pluto's surface.

A collision between Pluto and another KBO is believed to have created Charon, the largest of Pluto's five moons, in an event analogous to the impact that formed the Earth's Moon. Thus, Pluto is really a double KBO: Charon is half Pluto's size and one-ninth its mass. Every 6.4 days they orbit around a point that lies between them. After the Charon-forming impact, tidal heating should have melted a subsurface ocean on Pluto. Pluto probably has a rocky core that might supply enough radiogenic heat to maintain a subsurface ocean of ammonia-rich water today.

Organisms in Pluto's ocean would be limited by energy, so the biomass would be far smaller than on Europa. Any ocean is also probably at depths exceeding 350 km, and difficult to access in a future space mission, but life might be there.

Our consideration of Pluto shows that life might exist in remarkably unlikely places, and my list in Table 1 may be too conservative. Many of the cold outer Solar System objects used to be dismissed as potential habitats. But perceptions have shifted. It's possible that there are subsurface ammonia-rich oceans on Saturn's moon Rhea, Uranus's largest moons Titania and Oberon, and perhaps some large KBOs, such as Sedna, which is similar in size to Pluto yet 100 AU from the Sun. Given that so many objects in the Solar System are potential refuges for life, billions of similar possibilities surely exist throughout our galaxy.

Chapter 7
Far-off worlds, distant suns

The hunt for exoplanets

Beyond the Solar System, astronomers have discovered over 3,400 exoplanets, including candidates and confirmed bodies. It's easier to detect large ones, so nearly all are bigger than Earth and some are rather exotic. *Hot Jupiters* are Jupiter-size planets within 0.5 AU of their parent stars, some orbiting in only a few days. Then there are planets that sound like Earth on steroids, the *Super-Earths*. These are up to ten times the mass of our planet. Although harder to find, it's clearly just a matter of time before many Earth-sized exoplanets become known. How many will be habitable or even inhabited?

Of course, before identifying a *habitable* exoplanet, you have to find exoplanets. With the vast distances involved, the search is difficult but nonetheless astronomers have developed two classes of methods. The first, *indirect detection*, looks for stellar properties, such as position or brightness, which are affected by the presence of unseen planets. The second is *direct detection* of a planet with an image or a spectrum of its light.

The four indirect detection methods are: astrometry, stellar Doppler shift, transits, and gravitational microlensing. The key to the first two is that a planet and star orbit a common centre of mass. For example, if a planet and star had exactly the same mass, they would orbit a point halfway between them. In reality, a planet is smaller than a star, so the centre of mass is closer to the star and perhaps inside it. But in all cases, the star will follow a little orbit around the centre of mass and 'wobble' even when its planets can't be seen. In our own Solar System, Jupiter's twelve-year orbit causes twelve-year wobbles of the Sun's position. Saturn adds in another smaller twenty-nine-year wobble, corresponding to its twenty-nine-year orbit around the Sun.

Astrometry measures the motion of a star in the sky using telescopes. It's sensitive to big planets far away from their star so it could be used to find planetary systems similar to our own. But you have to wait many years or decades to track the effects of planets far from their star.

The second technique, the *Doppler shift* or *radial velocity* method, relies on the fact that when the light of a star is split into all its colours, the spectrum has dark bands like a bar code. The elements in a stellar atmosphere absorb photons coming from the star's interior, which cause the dark lines. If the star moves toward or away from the Earth, the lines shift to higher (bluer) or lower (redder) frequencies, respectively. Everyone has experienced the *Doppler effect* in sound. When a wailing police car approaches, the sound waves are bunched into a high-frequency squeal, but after the car passes by, they become a low-frequency drone. A similar frequency shift occurs with light. A rhythmic red shift and blue shift of lines in a stellar spectrum shows that the star is wobbling and has a planet around it. The size of the shift indicates the mass of a planet, while the pacing gives the time for the planet to complete an orbit.

In 1995, Didier Queloz (then a student) and his mentor, Michel Mayor, from the University of Geneva, detected the first planet around a Sun-like star using Doppler shift. It was a planet of at least half a Jupiter mass orbiting a star in the constellation of

Pegasus every four days. It was a total surprise. No one expected a giant planet so close (0.05 AU) to its parent star because giant planets shouldn't form there. Now we understand that some extrasolar planets undergo planetary migration early in their history (Chapter 3) and we end up seeing what survived the jostling. In 2012, Doppler data suggested a possible Earth-mass planet just 4.3 light years away, orbiting the star Alpha-Centauri B. The planet, if present, lies 0.04 AU from Alpha-Centauri B, which is so close that its surface is probably molten rock.

One subtlety with the Doppler shift technique is how we view the planetary system. If the orbit is face-on (described as an inclination of zero degrees from a plane perpendicular to the line of sight), there's no to-and-fro Doppler shift. The more edge-on the orbit, the greater the Doppler shift. It becomes maximal at an edge-on inclination of 90 degrees. There's often no way of knowing the inclination, so a measured Doppler shift might be smaller than the ideal edge-on case, allowing us only to infer a *minimum* planetary mass. The Doppler technique is most sensitive to big planets close to their star, so that's why most of the exoplanets initially detected were hot Jupiters. But we now know from the transit method that hot Jupiters are actually only a tiny minority of exoplanets, around 0.5 to 1 per cent.

The *transit* method measures the decrease in starlight when a planet crosses the face of a star, which can happen if we're lucky enough to view an exoplanet system virtually edge-on. Such geometry is statistically rare, but there are so many stars that if you gaze widely and long enough, transits will be seen. From the dip in starlight, it's possible to determine a planet's diameter if the star's size can be estimated. In turn, the planet's orbital period and distance from its star can be calculated from the cycling of the dimming. NASA's *Kepler* mission is a telescope that operated from 2009 to 2013 from an orbit around the Sun staring continuously at 145,000 main sequence stars in the constellation of Cygnus (the Swan) to look for transits where most of the stars are

500–3,000 light years away. *Kepler* has found over 3,200 exoplanet candidates, which are confirmed by checking for periodic dimming or using another detection technique, such as Doppler shift. In 2017, the Transiting Exoplanet Survey Satellite (TESS) telescope will begin examining two million stars for transits.

The fourth indirect method is microlensing. When an object passes between a distant star and us, Einstein's Theory of General Relativity predicts the bending of light by the object's gravitational field. The foreground object effectively acts as a lens, focusing the light and making the distant star appear gradually brighter. But if a planet orbits a lens star, the background star can brighten more than once. This sensitive technique can find Earth-mass planets at orbital distances of 1.5–4 AU around a star. Unfortunately, once the alignment has happened, it's virtually impossible to follow up with more detailed measurements because the objects are typically very distant. However, microlensing can gather statistics. Large planets drifting alone in interstellar space can also act as lens objects, and one result is that there appears to be many unbound Jupiter-mass planets floating between the stars. These lonely worlds were presumably ejected from their extrasolar systems.

For astrobiology, the most important techniques are *direct detections* that capture light from a planet. Direct detection is challenging because a planet is a dim body close to a vastly brighter star. Nonetheless, telescopes in space, such as the Hubble Space Telescope, and extremely large telescopes on the ground have accomplished direct detection using a *coronograph*, which is a mask to block the starlight. The term 'coronograph' comes from techniques that were originally used to block out sunlight to study the Sun's corona, which is the wispy halo seen in a solar eclipse. Ground-based telescopes also use *adaptive optics*, which are procedures to sense and correct the distortions caused by the shimmering of the Earth's atmosphere.

A space-based telescope has the advantage of avoiding the blurs from the Earth's atmosphere, but there's the question of which part of the spectrum to examine. In general, planets emit mostly infrared radiation and reflect visible starlight. If we looked at our Solar System from afar, the Sun, being hot and big, would outshine the Earth by a factor of about ten million in the infrared and ten billion in visible light. So looking for a distant Earth in the infrared gives a thousand times better contrast than in visible light. Unfortunately, light also spreads and blurs (diffracts) when it encounters any object such as a telescope and this is worse with infrared light than with visible. Fortunately, a technique called interferometry can cancel unwanted light. Noise-cancelling headphones use the same principle to silence undesirable sound waves.

Light waves have crests and troughs like water waves. *Nulling interferometry* uses more than one telescope mirror to precisely align light waves arriving from a point in the sky so that wave crests cancel troughs and light is turned into darkness. In this way, starlight can be 'nulled' to see planets. Both NASA and the European Space Agency have studied space telescopes, called *Terrestrial Planet Finder* and *Darwin*, respectively, which use interferometry to capture images and spectra of exoplanets around nearby Sun-like stars.

The results of exoplanet surveys to date generally show that the number of exoplanets increases at lower masses, so rocky, Earth-sized planets should be common. Density estimates for some exoplanets also allow assessment of composition. Density has been inferred using mass from the Doppler method and size from transit methods. Also, if a transiting Earth-sized planet has at least one Neptune-like sibling, a method called *transit timing variations* can provide the mass of planets and thus density. Even if the larger planet doesn't transit its parent star because of its orbital inclination, it can cause slight variations in the smaller planet's transit times that allow planetary masses to be calculated. With density measurements, there are increasing clues about which exoplanets are made of gas, rock, water, or some mixture.

The habitable zone

When we were considering life in the Solar System, we concentrated on bodies with liquid water because we're sure that's a requirement for Earth-like life, and the same idea applies to exoplanets. Whereas dry planets with subsurface oceans are worth investigating with space probes within our Solar System, they generally wouldn't be targets for exoplanet astrobiology. There's no prospect in the near future of sending spacecraft to visit exoplanets, so everything must be done with telescopes looking at light from afar. Consequently, we care most about exoplanets with liquid water right at the surface. These have the chance of a biosphere that pumps lots of gases into an atmosphere. In principle, such biogenic gases are detectable in the light from the planet and might indicate life.

In 1853, William Whewell noted that Earth's distance from the Sun allowed liquid water between what he called a 'central torrid zone' and an 'external frigid zone'. In the 1950s, the American astronomer Harlow Shapley (1885–1972) (who discovered the dimensions of our galaxy) also talked about a zone around stars where planets could have liquid water on their surface. If a planet is too far away from its host star it ices over, and if it's too close it's too hot for liquid water. Nowadays, the term *habitable zone* (HZ) refers to the region around a star in which an Earth-like planet could maintain liquid water on its surface at some instant in time. We specify a particular time because stars age and brighten or dim, so the HZ moves. In contrast to the HZ, the *continuously habitable zone* (CHZ) is the region around a star in which a planet could remain habitable for some specified period, usually the star's main sequence lifetime.

The width of the HZ around different types of main sequence stars, including the Sun, has been estimated. The inner edge is set by a planet's susceptibility to the runaway greenhouse effect, while the outer boundary is usually determined by the failure of greenhouse warming when carbon dioxide is cold enough to condense into clouds of dry ice or dense enough to scatter away sunlight. The latter was a potential problem for early Mars (Chapter 6). For the Sun, the HZ's inner edge is around 0.85–0.97 AU, while the periphery is 1.4–1.7 AU. The spread reflects the uncertain effects of water and carbon dioxide clouds at the inner and outer borders, respectively. For example, clouds might cool a close-in planet by reflecting sunlight, so the inner edge might be 0.85 instead of 0.97 AU. For the most optimistic outer boundary, Mars resides within the HZ. However, Mars's small size has led to a thin atmosphere that can't sustain liquid water. In other words, if Mars had been as big as the Earth, perhaps it might be habitable today. So planet size matters and the HZ is not the whole story for habitability but just a first guide.

Around other stars, the HZ is closer in or further out depending on whether a star is cooler or hotter than the Sun. For example, a K star, which is cooler (Fig. 1), would have a tight habitable zone from about the orbit of Mercury to that of the Earth. An even cooler M-type red dwarf would have the centre of its habitable zone at only 0.1 AU or so, well within the 0.4 AU orbit of Mercury.

In fact, the HZ of faint M dwarfs is so snug that planets will experience *tidal locking*, which occurs when gravity sets the rotation period. The same phenomenon makes one face of the Moon point toward the Earth. The Moon is tidally locked and spins on its axis once for every orbit around the Earth. The match between orbital and rotation periods, called *synchronous rotation*, is no coincidence. The Moon elongates slightly in the Earth–Moon direction, and if this orientation deviated, the Earth's gravity would twist the Moon's skewed bulges back into alignment.

A synchronously rotating planet in the HZ of an M dwarf would have one side in sunlight and the other in perpetual darkness. However, this doesn't make such planets uninhabitable. Although the night side would be cold, a moderately thick atmosphere can transport enough heat to warm the night side while cloud cover can cool the day side. Also, a large moon orbiting such a planet would have variable sunlight. So tidal locking in the habitable zones of M dwarfs is not a showstopper for life. That's encouraging because M dwarfs are the most common type of stars, making up about three-quarters of the total.

In recent years, assumptions defining the HZ have been questioned. Conventionally, the outer edge is determined by carbon dioxide condensation. But some large, rocky exoplanets might have thick hydrogen-rich atmospheres. We know from studying Titan and the giant planets that hydrogen and methane behave as greenhouse gases, and these condense at far colder temperatures than carbon dioxide. Also, on the inner edge, a water-poor planet with polar lakes rather than oceans might not succumb to a runaway greenhouse because it wouldn't have enough water to generate an atmosphere completely opaque to infrared radiation. Thus, the HZ might be wider than we think.

Is there a galactic habitable zone?

Some astronomers also argue that there's an optimal region in the galaxy for habitable planetary systems called the galactic habitable zone (GHZ). They note that the Sun is two-thirds of the way from the centre of the Milky Way, whereas planetary systems near the densely populated centre would be perturbed by supernovae or passing stars. At the other extreme, stars near the galaxy's edge are very poor in elements other than hydrogen or helium, and this might curtail planet formation. Astronomers have an odd convention (which sickens chemists!) of calling *all* elements other than hydrogen or helium 'metals'; the 'metal' content of a star is called its *metallicity*. Stars with giant exoplanets tend to be metal rich, and since giant exoplanets were the ones discovered first, it was initially thought that high metallicity was needed for planets to form in a nebula. More recently, no correlation with metallicity has been found in stars of Sun-like mass that have Super-Earths.

Another problem with the GHZ is that stars don't stay put. Recent research shows that stars wander across the galactic disc as a result of gravitational scattering by spiral arms. In any case, the exoplanets that will be scrutinized for signs of life in the foreseeable future will all be close, within a hundred light years of us. So whatever the validity of the GHZ, it's not a practical consideration for finding habitable planets any time soon.

Biosignatures, or how we find inhabited planets

Of course, the central question is whether we can find life. In 1990, NASA's *Voyager 1* spacecraft was heading out of the Solar System and looked back at the Earth from 6.1 billion km away. The famous picture it took is known as the 'Pale Blue Dot', in which Earth is so small that it's a single bluish pixel. If we knew nothing else, what could we deduce? If we said that the planet's colour was a mixture of blue oceans and white clouds that would just be guesswork. How can we do better?

Generally, to determine an exoplanet's properties and look for life, we need a planet's visible or infrared light spectrum. Different molecules and atoms absorb different frequencies in spectra. Thus, we can look for atmospheric gases such as oxygen or methane that can be produced by life. Such planetary features that indicate life are *biosignatures*. In fact, in the same year as *Voyager 1*'s 'Pale Blue Dot', the *Galileo* spacecraft, which was headed for Jupiter, obtained spectra of Earth. It was a sort of dry run for exoplanets and showed that you could detect Earth's atmospheric oxygen, methane, and abundant water. The simultaneous presence of oxygen and methane is evidence for life because these two gases should quickly react with each other unless a biosphere is producing great amounts of them, which prevents equilibrium. Thus, we say that Earth's atmosphere is in *chemical disequilibrium*. That's the kind of biosignature we would like from exoplanets.

Already, some information has been collected about exoplanet atmospheres by looking at differences in spectra when transiting exoplanets pass behind or in front of a star. The spectrum in the *primary transit* when a planet crosses the face of a star includes light passing through a ring of atmosphere around the planet. So subtracting the spectrum of just the star after the planet has passed by can isolate the spectrum of the planet's atmosphere. Alternatively, you take a spectrum when the planet is behind the star in its *secondary transit* and you subtract that from a spectrum of the planet plus star when the planet is beside the star. This produces the spectrum of the whole planet. In fact, this technique has been used to obtain infrared spectra of hot Jupiters with the Spitzer Space Telescope, which is a telescope orbiting the Sun.

With more sophisticated telescopes, we would like to determine an Earth-like exoplanet's surface temperature and whether it has liquid water. A spectrum might provide that, but you might only be able to see light from cloud tops or the upper atmosphere. In fact, Venus is veiled in this way. For exoplanets that are similarly obscured, we'll need to infer the amount of greenhouse gases from absorption lines in spectra and calculate a surface temperature. Together with an exoplanet's atmospheric composition, you could then infer whether liquid water is present. It might also be possible to look for an ocean through glint, which is the bright reflection spot on a smooth body of water at glancing angles. Finally, we should keep an open mind about *anti-biosignatures*. Microbial life will readily eat hydrogen or carbon monoxide gas, so an abundance of either of these might be considered an anti-biosignature for a planet in the habitable zone. Basically, anti-biosignatures say 'no one home'.

The search for extraterrestrial intelligence (SETI)

A different approach to finding life on exoplanets is to hypothesize that technological civilizations exist. If they do, we can look for their communications. This endeavour is the search for extraterrestrial intelligence (SETI) or more generally the search for *technosignatures*.

In 1959, Guiseppe Cocconi and Philip Morrison advocated looking for broadcasts of extraterrestrial civilizations using large radio telescopes. Others have since suggested watching for transmissions in visible light. There are practical reasons to look only for these signals. For example, X-rays are absorbed in the upper atmosphere, whereas radio and visible light are easy to generate and detectable across space. The main question for SETI is whether it's worth looking. Are there enough civilizations out there?

In 1961, Frank Drake (then at Cornell University) described a way to evaluate the potential number of transmitting civilizations in our galaxy. If we estimate the average number of civilizations that come 'on air' each year and their average lifetime, we can multiply those two quantities together to give the current number of transmitting civilizations. To give an analogy, if the number of new freshmen starting at a university is around 6,000 per year and they spend an average of four years on campus, the total undergraduate population at any one time is $6,000 \times 4 = 24,000$. Drake went further by devising six factors that determine the number of 'freshman' transmitting civilizations appearing per year. In full, we calculate N, the number of communicating civilizations, by multiplying the six numbers and the average lifetime, as follows:

N civilizations =

$$\underbrace{R_* \times f_{\text{planet}} \times n_{\text{habitable}} \times f_{\text{life}} \times f_{\text{intelligence}} \times f_{\text{civilizations}}}_{\text{number of communicating civilizations appearing each year}} \times \underbrace{l}_{\text{their average lifetime}}$$

This is the famous *Drake Equation*. The first number R_*, is the birth rate of stars suitable for hosting life. Astronomers observe about ten new stars per year of types G, K, and M. All the other factors are as follows:

f_{planet} is the fraction of such stars having planets;

$n_{habitable}$ is the average number of planets per planetary system that are habitable;

f_{life} is the fraction of those planets on which life originated and evolved;

$f_{intelligence}$ is the fraction of inhabited worlds that developed intelligent life;

$f_{civilizations}$ is the fraction of those worlds that developed civilizations capable of interstellar communication;

and L is the lifetime of those communicating civilizations

Several factors in the equation are unknown. But, what the heck! Let's use the Drake Equation anyway. Current exoplanet searches suggest that at least two-thirds of all stars have planets, so let's take f_{planet} as 2/3. Data from the *Kepler* mission are still being analysed, but suggest at least that 1 in 100 of planets are habitable, so let's set $n_{habitable}$ = 1/100. We will simply have to guess all other parameters. Life developed rapidly on Earth, so let's presume that life originates on half of the habitable planets, i.e. f_{life} = 1/2. Let's suppose that the fraction of biospheres that develop intelligence is $f_{intelligence}$ = 1/8. Let's also guess that one in ten intelligent biospheres develop civilizations capable of interstellar transmissions, so $f_{civilizations}$ is 1/10. Finally, the lifetime, L, of communicating civilizations is sociological speculation, but let's say 10,000 years. So how many communicating civilizations in the Milky Way do we get? The answer is four, from $10 \times (2/3) \times (1/100) \times (1/2) \times (1/8) \times (1/10) \times 10,000$. Fine. But can we constrain the probabilities better?

Exoplanet studies will eventually nail down the second and third terms in the Drake Equation and might have something to say about the probability of life arising, f_{life}, if biosignatures are detected. Optimists think that a planet with liquid water and the right materials develops life easily. At present, we don't really know.

The next term concerns intelligence, so how probable is that? A relevant issue might be that only some biological solutions work to solve specific problems. In zoology, we often find the same trait shared by organisms in unrelated lineages, a result of what is called *convergent evolution*. Convergence occurs for specific functions and ecological niches. For seeing, eyes have evolved in at least forty different animal groups. A dog's life, of all things, is an example of a convergent niche. The Tasmanian wolf (a marsupial) evolved dog-like traits similar to the Mexican wolf (a placental mammal). Evolutionary convergence is so common that it means that organisms with legs, pairs of eyes, and certain ecologies might be inevitable because only these are physical solutions to the problems of walking, stereo vision, and occupying particular niches.

Whether technological intelligence is unique or convergent is contentious. On Earth, it was slow to appear. It took four billion years to build up sufficient O_2 to allow animal life. Then another few hundred million years passed before technological life. Dinosaurs reigned for about 170 million years and yet we find no sign of technology, neither fossil tools nor dinosaur microwave ovens. On the other hand, there are some lineages where brains appear to have had value and grown. Brain mass itself is a poor measure of intelligence because a big body often needs a big brain to run it. Instead, the Encephalization Quotient (EQ) is used, which is the ratio of the brain mass of an animal to the brain mass of an average animal of the same body mass. Thus, an average animal has an EQ of 1. Scores higher than 1 are brainy and scores below 1 more brainless (Fig. 11). Humans and chimps have EQs of about 7.4 and 2.5, respectively, which means that they have larger brains than expected by these factors. A rabbit has an EQ of 0.4, which would please Elmer Fudd. As Aristotle noted, 'man has the largest brain in proportion to his size'.

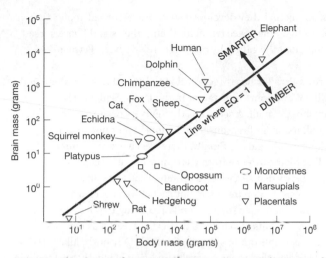

11. Brain and body mass for some different mammals. Animals that plot to the left of the diagonal line have more brain mass than typical

Palaeontologists have discovered an increase in EQ in certain lineages over time, most notably the genus *Homo*. The EQ of *Homo habilis*, a hominin that lived about 2 Ma, was only 4, for example. In toothed whales (dolphins, sperm whales, and orcas), EQ increased around 35 Ma, when they developed echolocation to find fish and friends. 'Friends' may be the key. Generally, intelligence is larger in social animals, such as 5.3 for a dolphin. Intelligence probably aids survival and getting a mate, which, in turn, promotes more offspring. However, there's much debate about which evolutionary pressures are behind intelligence.

A final question about SETI is called the *Fermi Paradox*, and was conceived by Enrico Fermi (1901–54), the Nobel Prize-winning physicist who built the world's first nuclear reactor in 1942. His idea arose from a lunchtime conversation. Stars in the Milky Way disc date from nine billion years ago, whereas the Earth is only 4.5 billion years old. Thus, if intelligent life is common, many technological civilizations should have arisen long before us.

Assuming that they developed space travel (or self-replicating robots to do the space travel for them), they should have spread throughout the galaxy by now. 'Where are they?' Fermi asked.

Fermi's Paradox has three solutions. One is that we're alone in the Milky Way because one factor in the Drake Equation is vanishingly small. A second is that Fermi's premise is incorrect. For all manner of reasons, civilizations might not go around colonizing the galaxy. Finally, there are civilizations but they're hiding their existence from us (or most of us). Scientists particularly dislike the third option because it's an ad hoc hypothesis that's untestable, but it's beloved of science fiction writers, tabloid newspapers, and (according to polls) one-third of American adults who believe that extraterrestrials have visited. The reasonable options are the first two. Obviously, all efforts to find life elsewhere bear upon the question of whether we're alone and the first solution. If SETI succeeds, we might also gain insight into other civilizations, if we could understand their signals. At the moment, the only known life is here, so, like Fermi, we can only wonder, 'Where are they?'

Chapter 8
Controversies and prospects

The Rare Earth Hypothesis

The big, unanswered questions of astrobiology generate controversy. One debate surfaced in 2000 when Peter Ward and Don Brownlee, my colleagues at the University of Washington in Seattle, published a bestselling book, *Rare Earth*. In essence, their *Rare Earth Hypothesis* is that the fortuitous circumstances that have allowed complex life on the Earth are so uncommon that Earth might harbour the only intelligent life in the Milky Way. Amongst their arguments were the good fortune of Earth being in the right place in the galaxy, having Jupiter in our Solar System to capture comets that might otherwise collide with the Earth, Earth's unusual recycling of volatiles by plate tectonics to keep the atmosphere going, the contingencies in obtaining an oxygen-rich atmosphere, and the luck of having a large Moon that stabilizes the Earth's axial tilt and so its climate.

Rare Earth was a polemic that railed against the *Copernican Principle*. The latter idea (named after Nicolas Copernicus, whose Sun-centred system knocked the Earth from its perceived place as the centre of the universe) holds that there's nothing special about our location. Astronomers note several factors in its favour. First, the Earth is surely one of many rocky planets in the universe. Furthermore, our Sun, a G-type star, is not special because around one in ten stars are G type. We also live in a humdrum location in the galaxy, along one of many spiral arms. Finally, our galaxy is unremarkable among many in the observable universe. As Stephen Hawking has put it, 'The human race is just a chemical scum on a moderate-sized planet, orbiting around a very average star in the outer suburb of one among a hundred billion galaxies.'

Advocates of the Copernican Principle also note how the great advances of science highlight the folly of assuming that we're special. We need think only of Galileo's *The Starry Messenger* (1610), which reported his observations of moons of Jupiter and the surface of the Moon, showing that these bodies were not heavenly perfections as supposed by theologian philosophers but explained by the same physics that we have on Earth. Similarly, Darwin's *Origin of Species* (1859) overturned the conceit that humans exist outside the rest of biology.

The problem with the Rare Earth Hypothesis is that it assumes too much knowledge about habitability, whereas, in reality, much is uncertain. Recently, for example, it's been discovered that the wandering of stars means that the galactic habitable zone is a less concrete concept than previously thought. The results of NASA's *Kepler* mission also show that planets around other stars are common, including plenty of Jupiter-size bodies. The argument that having a Jupiter-sized planet in just the right place lowers the rate at which comets or asteroids hit the Earth has also been challenged. Certainly, Jupiter mops up some impactors like a cosmic vacuum cleaner. In 1994, the comet Shoemaker–Levy 9 smashed into Jupiter, while in 2009 and 2010, the scars of further impacts were seen on Jupiter. It's well established that Jupiter helps to deflect and eject comets that come from a halo of such icy objects, the Oort Cloud, which surrounds the Solar System at about 50,000 AU distance. However, near-Earth asteroids and comets from within the plane of the Solar System represent more than 75 per cent of Earth's impact threat and Jupiter can actually destabilize those objects. So Jupiter is two-faced. Some calculations suggest that Jupiter is actually a net foe rather than friend.

Other arguments for Rare Earth are also ambiguous. In the Solar System, Earth uniquely has plate tectonics. However, for plate tectonics, a planet must be big enough to have adequate internal heat to drive the plate motion and it probably needs seawater to cool oceanic plates and lubricate their movement. In the Solar System, only the Earth qualifies. But that doesn't mean that Earth-like exoplanets in habitable zones might not also be suitable for similar tectonics.

While the Rare Earth Hypothesis is correct that planets without a large moon will suffer larger axial tilt variations than Earth, climatic variations at low latitudes might be benign. In fact, thicker atmospheres, more extensive oceans, and lower rotation rates of an exoplanet can smooth the climatic differences between pole and tropics caused by a varying tilt. Finally, the question of oxygen not accumulating on other Earth-like planets might go in the other direction to that assumed in the Rare Earth Hypothesis. Some planets might be more favourable for oxygen-rich atmospheres than Earth because their volcanoes pump out a smaller proportion of gases that react with oxygen. Oxygen might build up more easily.

What does seem to be correct about the Rare Earth Hypothesis is that microbial-like life should be much more common than intelligent life. Microbes have a remarkable range of metabolisms and can live in a far wider variety of environments than complex organisms. But any definitive statements about the prevalence of complex life—one way or the other—simply lack data to support them and we should be sceptical. As Carl Sagan famously remarked, 'It pays to keep an open mind, but not so open that your brains fall out.'

Prospects for astrobiology and finding life elsewhere

The excitement of astrobiology is that it tries to answer questions such as the origin of life and whether we're alone in the universe. With advances in technology, it's increasingly likely that major discoveries will be made in the coming decades.

In the area of understanding early life, it's likely that realistic self-replicating genomes will be made in the lab. This would provide great insights into the origin of life. Several groups are studying the RNA World or its variants. There are also projects to drill deeply into old sedimentary rocks in South Africa and Australia, which will surely make new discoveries about the earliest life and environment.

In the Solar System, Enceladus ought to be one of the highest priorities for the world's space agencies. Enceladus has a source of energy (tidal heating), organic material, and liquid water. That's a textbook-like list of those properties needed for life. Moreover, nature has provided astrobiologists with the ultimate free lunch: jets that spurt Enceladus's organic material into space. Technology certainly exists to build a spacecraft to swing by Enceladus and sample the organics in the jets. Better still, the material could be returned to Earth for analysis.

In fact, spacecraft to collect extraterrestrial samples and return them to Earth, which are *sample return* missions, are the future in understanding the history of Mars and Venus and whether either of these planets was once inhabited. A sample return mission for Venus is in the distant future, but one for Mars is a strategic goal of NASA's and ESA's current programmes.

Around Jupiter, Europa is probably the best prospect for life. The first step would be a Europa orbiter to study the moon in detail and determine the thickness of the ice above a subsurface ocean or lake lenses. The next steps might involve landers, and possibly robots to melt through the ice using radioactive heat generators. Eventually one imagines submarines diving though a Europan ocean.

Apart from Enceladus, Titan is a target of astrobiology amongst Saturn's moons. A huge scientific leap could be made if a lake lander—a sort of interplanetary boat—could float on the lakes in the polar regions of Titan and find out which substances make up the organic liquids. Furthermore, a Titan orbiter could do the kind of reconnaissance that would determine the depth of Titan's subsurface ocean and study Titan's surface.

One certainty for the future is that exoplanet discoveries will continue to spur an interest in astrobiology. I anticipate the discovery of many Earth-like planets inside the habitable zone of other stars, dead planets with almost pure carbon dioxide atmospheres, water worlds covered entirely in glinting oceans, and young Venus-like planets sweating off their oceans into space from runaway greenhouse effects.

When astrobiology came to the fore as a discipline in the 1990s, some questioned its future and wondered if it might be a fad that fades, perhaps because of disappointment in not quickly finding extraterrestrial life or a failure to answer questions about life's origin. However, the discovery of Earth-sized exoplanets in habitable zones will ensure that the possibility of life elsewhere becomes more relevant than ever. Astrobiology is here to stay.

Further reading

Chapter 1: What is astrobiology?

Much lengthier introductions to astrobiology are found in the following textbooks:

J. O. Bennett, G. S. Shostak. *Life in the Universe.* (San Francisco: Pearson Addison-Wesley, 2012).

D. A. Rothery et al. *An Introduction to Astrobiology.* (Cambridge: Cambridge University Press, 2011).

K. W. Plaxco, M. Gross. *Astrobiology: A Brief Introduction.* (Baltimore: Johns Hopkins University Press, 2011).

J. I. Lunine. *Astrobiology: A Multidisciplinary Approach.* (San Franciso: Pearson Addison Wesley, 2005).

W. T. Sullivan, J. A. Baross (eds). *Planets and Life: The Emerging Science of Astrobiology.* (Cambridge: Cambridge University Press, 2007).

The development of astrobiology from the 1950s onwards is described by: S. J. Dick, J. E. Strick. *The Living Universe: NASA and the Development of Astrobiology.* (New Brunswick, NJ: Rutgers University Press, 2004).

An old classic on the nature of life is: E. Schrödinger. *What Is Life?* (1944; Cambridge: Cambridge University Press, 2012).

Chapter 2: From stardust to planets, the abodes for life

A readable discussion of modern Big Bang theory is given by:

C. Lineweaver, T. Davis. 2005. Misconceptions about the Big Bang. *Scientific American* 292: 36–45.

A popular account of Arthur Holmes's quest to find the age of the Earth is: C. Lewis. *The Dating Game: One Man's Search for the Age of the Earth*. (Cambridge: Cambridge University Press, 2012).

Chapter 3: Origins of life and environment

The science of the origin of life is described by: R. M. Hazen. *Genesis: The Scientific Quest for Life's Origin*. (Washington, DC: Joseph Henry Press, 2005).

A lucid description of the early evolution of life on Earth is: A. H. Knoll. *Life on a Young Planet: The First Three Billion Years of Evolution on Earth*. (Princeton: Princeton University Press, 2003).

Chapter 4: From slime to the sublime

The Earth's formation, evolution, and habitability are covered in: C. H. Langmuir, W. S. Broecker. *How to Build a Habitable Planet: The Story of Earth from the Big Bang to Humankind*. (Princeton: Princeton University Press, 2012).

Chapter 5: Life: a genome's way of making more and fitter genomes

A widely used introductory textbook on modern microbiology is: M. T. Madigan et al. *Brock Biology of Microorganisms*. (San Francisco: Benjamin Cummings, 2012).

The effects of life on the Earth's chemistry on a global level are described in the following textbook: W. H. Schlesinger, E. S. Bernhardt. *Biogeochemistry, Third Edition: An Analysis of Global Change*. (San Diego: Academic Press, 2013).

Chapter 6: Life in the Solar System

The planets of the Solar System and their habitability are described in the following textbook: J. J. Lissauer, I. de Pater. *Fundamental Planetary Science: Physics, Chemistry and Habitability*. (Cambridge: Cambridge University Press, 2013).

Chapter 7: Far-off worlds, distant suns

A readable book that discusses the search for habitable exoplanets is:
J. F. Kasting. *How to Find a Habitable Planet*. (Princeton:
Princeton University Press, 2010).

Chapter 8: Controversies and prospects

The controversial but engrossing book that argues for the scarcity of
complex life is: P. D. Ward, D. Brownlee. *Rare Earth: Why
Complex Life is Uncommon in the Universe*. (New York:
Copernicus, 2000).

Index

SOCIAL MEDIA
Very Short Introduction

Join our community

www.oup.com/vsi

- Join us online at the official Very Short Introductions **Facebook** page.
- Access the thoughts and musings of our authors with our online **blog**.
- Sign up for our monthly **e-newsletter** to receive information on all new titles publishing that month.
- Browse the full range of Very Short Introductions online.
- Read **extracts** from the Introductions for free.
- Visit our library of **Reading Guides**. These guides, written by our expert authors will help you to question again, why you think what you think.
- If you are a teacher or lecturer you can order inspection copies quickly and simply via our website.

Visit the Very Short Introductions website to access all this and more for free.
www.oup.com/vsi

GALAXIES
A Very Short Introduction
John Gribbin

Galaxies are the building blocks of the Universe: standing
like islands in space, each is made up of many hundreds of
millions of stars in which the chemical elements are made, around
which planets form, and where on at least one of those planets
intelligent life has emerged. In this *Very Short Introduction*,
renowned science writer John Gribbin describes the
extraordinary things that astronomers are learning about
galaxies, and explains how this can shed light on the origins
and structure of the Universe.

www.oup.com/vsi

PLANETS
A Very Short Introduction
David A. Rothery

This *Very Short Introduction* looks deep into space and describes the worlds that make up our Solar System: terrestrial planets, giant planets, dwarf planets and various other objects such as satellites (moons), asteroids and Trans-Neptunian objects. It considers how our knowledge has advanced over the centuries, and how it has expanded at a growing rate in recent years. David A. Rothery gives an overview of the origin, nature, and evolution of our Solar System, including the controversial issues of what qualifies as a planet, and what conditions are required for a planetary body to be habitable by life. He looks at rocky planets and the Moon, giant planets and their satellites, and how the surfaces have been sculpted by geology, weather, and impacts.

STARS
A VERY SHORT INTRODUCTION
Andrew King

Every atom of our bodies has been part of a star. Our very own star, the Sun, is crucial to the development and sustainability of life on Earth. This *Very Short Introduction* presents a modern, authoritative examination of how stars live, producing all the chemical elements beyond helium, and how they die, sometimes spectacularly, to end as remnants such as black holes.

Andrew King shows how understanding the stars is key to understanding the galaxies they inhabit, and thus the history of our entire Universe, as well as the existence of planets like our own. King presents a fascinating exploration of the science of stars, from the mechanisms that allow stars to form and the processes that allow them to shine, as well as the results of their inevitable death.